SOCIETY AND FREEDOM

Nelson-Hall Series in Sociology

Consulting Editor: Jonathan H. Turner
University of California, Riverside

SOCIETY AND FREEDOM

An Introduction to Humanist Sociology

Second Edition

Joseph A. Scimecca
George Mason University

Nelson-Hall Publishers/Chicago

Copy Editor: Dorothy Anderson
Typesetter: Precision Typographers
Printer: Capital City Press
Cover Painting: "Paris" by Susan Rodwan

Library of Congress Cataloging-in-Publication Data

Scimecca, Joseph A.
    Society and freedom : an introduction to humanist sociology /
Joseph A. Scimecca. — 2nd ed.
        p.   cm.
    Includes bibliographical references and index.
    ISBN 0-8304-1376-6
    1. Sociology. 2. Liberty. 3. Humanism.   I. Title.
HM51.S356   1995   94-17698
301—dc20   CIP

Manufactured in the United States of America

10 9 8 7 6 5 4 3 2 1

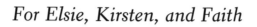

For Elsie, Kirsten, and Faith

CONTENTS

Preface to the Second Edition

Writing the second edition of a book should be much easier than writing the first edition. The conceptual framework has been laid out and theoretically all one has to do is update. Yet, anyone who has gone through the process knows that it is not that simple. In the years since 1981, when *Society and Freedom* was first published, American society has changed, the discipline of sociology has changed, and I have changed. In light of these changes, some who read the first edition may say I have become more conservative, and others will say I have become more radical. Both groups I think would be right.

Yet *Society and Freedom,* then and now, is different from other texts—it takes sides. It not only surveys the basic concepts of sociology from a humanist point of view, it also contends that dispassionate observation and analysis are not enough. As sociologists, we must actively challenge views and conditions that restrain human potential. As a humanist sociologist, I believe that the study of society begins with the premise that humans are free to create their social world and that whatever impinges upon that freedom is ultimately negative and destructive. This is the primary tenet of a humanist sociology.

By stating my opinions and not hiding behind a value-neutrality that I do not believe in, I invite readers to examine their own beliefs. In the years of teaching introductory sociology, I have found that this approach leads not only to lively discussion but also to a greater understanding of social behavior.

Instructors who emphasize a humanist approach may use *Society and Freedom* as a primary text. I have introduced and integrated into the discussions the major sociological concepts: society, culture, roles, social stratification and inequality, institutions, legitimacy, and authority. Instructors offering more traditional courses who wish to expose their students to the humanist perspective will find that care has been taken to make the text a convenient, and I hope rewarding, supplement.

In writing both editions to this text, I have incurred many debts. Primarily, I would cite the late C. Wright Mills, Ernest Becker, Alfred McClung Lee, and Richard Quinney, who was my teacher and now is a friend. Any intellectual framework I have I owe to these four social scientists. Thanks to the many readers who kept the first edition of *Society and Freedom* in print for eight years. Thanks again to my colleagues in the Department of Sociology and Anthropology at George Mason University, a department that proves daily that people do not have to agree ideologically to work well together. I am a better sociologist for having been a member of this department. Special thanks to Glen Goodwin and Jill Bystydzienski, who read and commented on the manuscript. It goes without saying that much of the strength and none of the weaknesses of this book are attributable to them. Finally, thanks to the women in my life, my wife Elsie, and my daughters Kirsten and Faith. They make my life worthwhile.

One last note. Throughout the book I have tried to use nonsexist pronouns. However, I have not changed the words of other authors cited or used *sic*. This would be unfair to them.

CHAPTER 1

Sociological Perspectives and Humanist Thought

Like traditional or mainstream sociology, humanist sociology is a perspective, a way of perceiving the world. Just as a Freudian psychologist focuses on sexual drives, defense mechanisms, and the Oedipal complex, or an economist looks at the production and distribution of goods and services to explain human behavior, sociologists are concerned primarily with the *social structure* of society (the groups people belong to and the positions they occupy in them) and its relationship to the personality of the individual. Unlike traditional sociologists, who often accumulate knowledge for knowledge's sake, humanist sociologists seek to *use* the knowledge they uncover to benefit people.

Foremost among the concerns of humanist sociologists and the principles that underlie the sociology presented in this text is the insurance and maximization of freedom for the individual. To be free is to be able to choose among alternatives. The word *freedom* will be used to mean the *maximization of alternatives.* Maximization implies that freedom is never total or absolute but is always a matter of "more or less"—the more limitations, the less freedom—therefore, humanist sociology is concerned with these limitations of behavior. The late anthropologist Ernest Becker wrote:

> When we ask what caused things to develop as they are now, how man
> in society got to be as he is, the only relevant principle must be the principle
> of human freedom; the only possible synthetic framework must be one
> that explains differences in human freedom in society and history.[1]

The formal definition of humanist sociology used here is: *The study of human freedom and of all the social obstacles that must be overcome*

in order to insure this freedom.[2] Not all those who consider themselves
humanist sociologists would accept this definition. Others offer different
meanings of humanist sociology.[3] However, a sociology grounded in the
principle of freedom (the *maximization of alternatives*) is a solid base
from which to clarify this alternative sociology. What follows, then, is a
view that is a synthesis of varied opinion, one that can provide a framework for those who may feel themselves outside the boundaries of mainstream sociology. The sociology put forth here is a discipline of thought
in which the dignity, interests, and values of human beings are of primary importance.

What is it in society that diminishes human dignity? Why do people
blindly follow leaders who constantly deceive them? Why is it that some
people have so much more than others and that those who have less
accept this condition? Who has power and how is this power used? How
can a society based on freedom, justice, and equality be achieved? These
are some of the questions that arise from the concept of freedom used
here and for which answers are sought. In short, a sociology that has
not lost sight of the Enlightenment principle of seeking to liberate the
human spirit and ensure the progressive development of the person is
the principle that guides the quest for a humanist sociology.

With this in mind, major perspectives in sociology—structural-
functionalism, conflict theory, and symbolic interactionism—are examined in light of a humanist orientation. The first perspective to be looked
at is structural-functionalism, or just functionalism.

Functionalism

The main principle of the functionalist approach is that everything
within a society contributes to the maintenance or support of the society.
Sociologists who accept this means of understanding human behavior
extend this principle to claim that every element within every subsystem
in the society serves a function for the continuation of the larger unit
of which it is a part. For example, expectations between a husband and
wife are seen as a function of maintaining the family; the family, in turn,
produces and teaches children how to be functioning members of the
larger society.

August Comte

Auguste Comte (1798–1857), the father of sociology, who coined
the word *sociology* in the 1830s, was among the first to look at human
behavior from the functionalist perspective. Comte likened society to
an organism, viewing it as a functionally organized system in which all

the elements were in harmony. If this harmony was disrupted, the system or society would keep changing until a balance was restored.[4]

Comte, following the lead of St. Simon, an early socialist and well-known philosopher of the time for whom Comte worked as a secretary, added the perspective of "positivism," or scienticism, to functionalism. Civilization was at the beginning of the age of science. Sociology would become the queen of the social sciences by reducing human phenomena—just as physical phenomena could be reduced—to general laws. Sociologists, beginning with what was known about the nature of the individual, would achieve an understanding of the predictable laws of society through observation and experimentation—through science. This view of sociology is still prominent today.

Herbert Spencer

The second important figure in the historical development of functionalism is Herbert Spencer (1820-1902), who contributed to the general, evolutionary theories prevalent in the nineteenth century. Darwin's idea of biological evolution—the progression from a lower order to a higher order—was used by Spencer to explain social phenomena. Sociology could thus become a science if it were based on the idea of evolutionary law. According to Spencer, there could "be no complete acceptance of sociology as a science so long as the belief in a social order not conforming to natural law survives."[5]

Like Comte, Spencer likened society to an organism. Government, the family, and organized religion all combine to play the same role in society as organs do in the body. They are to be studied both in terms of their evolutionary stages and the functions they serve at any given time.

Comte and Spencer were in general agreement on three of the major tenets of functional analysis: (1) the elements of a society are interrelated; (2) these elements function in society just like organs do in the human body; and (3) the proper means of discovering how these elements function is science.

Emile Durkheim

The next general classical figure instrumental in the development of functionalism is Emile Durkheim (1858-1917). Durkheim's concern throughout his life was the question: How is social order possible? This is the theme of his first important work, *The Division of Labor*,[6] in which he concentrated upon human solidarity, the feeling among people that they belong together. In this work, Durkheim analyzed two types of solidarity, *mechanical solidarity* (where people lived in small, uncomplicated, tightly bound societies) and *organic solidarity* (characterized by industrialization, urbanization, and modernization).

Durkheim's next work—and among his most influential—was *The Rules of Sociological Method*.[7] Here he eloquently put forth the argument that society has a reality of its own, that it is *sui generis*. In this view, society is an objective reality comparable to the reality of nature—something that is real and cannot be wished away. The explanation of social phenomena is not reducible to individual analysis but must be sought at the level of social facts. Social facts are characterized by two basic tenets; they are (1) external to and (2) constraining upon individuals. An example of a social fact used by Durkheim is suicide. In his book *Suicide*, Durkheim compared data from different religions, different periods of history, even different times of the day and year. He found that there were variations in the suicide rate within and between religions, between married and unmarried people, and the like. From these differences Durkheim concluded that suicide was a *social* and not an *individual* phenomenon. The reason people committed suicide was not because of individual distress (alcoholism, mental depression, and so forth), as was commonly believed at the time, but rather because of social pressures that promoted certain types of behaviors and inhibited others.[8]

Protestants, for example, are freer of the control of the church than are Roman Catholics. This lack of religious constraint produces a greater sense of individualism and hence greater feelings of despair, which can lead to an increased probability of suicide. Furthermore, Catholic doctrine provides specific injunctions against suicide. Catholics who commit suicide cannot be buried in consecrated ground. This leads to fewer suicides among Catholics than among Protestants. Similarly, when political or economic upheavals decrease the stability of a society, individuals are under less pressure from society's rules and are more likely to commit suicide. In short, for individuals who are more integrated into social groups, the probability is much less that they will commit suicide. Thus, suicide becomes a social fact, and social facts become the province of the sociologist.[9]

In *The Rules of Sociological Method*, Durkheim clearly established the basis of functional analysis, arguing that in order to achieve a complete explanation of a social fact, it is necessary to understand its function. According to Durkheim; "To explain a social fact it is not enough to show the cause on which it depends; we must also, at least in most cases, show its function in the establishment of social order."[10]

Durkheim's functionalism has been summarized by Lewis Coser:

> Whether he investigated religious phenomena or criminal acts, whether he desired to clarify the social act of the division of labor or of changes

in the authority structure of the family, Durkheim always shows himself a masterful functional analyst. He is not content merely to trace the historical origins of the phenomena under investigation, although he tries to do this also, but he moves from the search for efficient causes to inquiries into the consequences of phenomena for the structures in which they are variously embedded. Durkheim always thinks contextually rather than atomistically. As such he must be recognized as the direct ancestor of that type of functional analysis which came to dominate British Anthropology under the impact of Radcliffe-Brown and Malinowski; and which led, somewhat later, to American functionalism in sociology under Talcott Parsons and Robert Merton.[11]

It is to these two American sociologists, Parsons and Merton, who have set the directions of contemporary functional analysis, that we now turn.

Talcott Parsons

Talcott Parsons (1902–1979), perhaps more than anyone else, defined the boundaries of modern functional analysis. Parsons' primary interest was the social system, which he equated with society. This social system, or society, was to be analyzed in terms of what it needed to survive. Parsons held that the patterns that exist in society are there to enable the society to survive, and this need, in turn, explains what form the pattern will take.[12] Schools exist because some mechanism is needed to transmit skills to the next generation if a society is to survive; the pattern the schools take is the one that is best for what they do—educate the young. In short, societies create patterns (structures) that fulfill needs (functions) that maintain the society—hence the name structural-functionalism.

Parsons also held that a complimentarity exists between the needs of the society and the needs of the individuals who comprise the society—what is good for the society is good for the individual and vice-versa. Because Parsons did not see any real conflict between the needs of society and of individuals, he answered the question of how societies change by using the concepts of equilibrium and disequilibrium. Change occurs when systems become unbalanced and pass into a state of disequilibrium. This is an unnatural state, and the system must return to equilibrium for stability. For example, the student riots and demonstrations of the 1960s are explained by Parsons as a manifestation of disequilibrium, given that the next generation of students went back to academic pursuits, allowing society to eventually stabilize itself and return to equilibrium. Parsons believed that societies are characterized by extreme stability; change is the exception not the rule.

In order to better understand functionalism's concern with needs and change, the theories of Robert Merton (1910–), who studied under Talcott Parsons, will be examined.

Robert Merton

Robert Merton's contribution to functionalist thought is the revision of three basic assumptions that Parsons took for granted: (1) every item in a society is functional for the entire society; (2) every item in a society serves a positive function; and (3) every item in a society is indispensable.[13]

Merton modifies this stand by arguing that functional unity cannot be assumed. In any system there are dysfunctional items as well as functional alternatives, equivalents, or substitutes.[14]

Some elements may be functional for one group and not for another and, indeed, may ultimately become dysfunctional. The current controversy over abortion serves as an example of what Merton means. Although the support of right-to-life groups among Roman Catholics and fundamentalist Protestants may internally strengthen their religious groups, it brings them into conflict with pro-choice groups. The virulence of this conflict can be readily seen as dysfunctional to the political stability of the society. And many a politician caught between two emotionally charged constituencies would question the indispensability of pro-life or pro-choice groups.

For Merton, the questions of what is functional and for whom must be open to investigation by the functionalist sociologist and extends Parsons' argument. Whether or not this expanded view of functionalism holds any promise for humanist sociology is taken up after we look at the more recent developments in functionalist theory.

Recent Developments in Functionalist Theory

Jeffrey Alexander argues that an important change occurred during the 1980s in the development of functionalism.[15] Parsons and Merton had dominated the field from around World War II through the 1960s. During the 1970s, conflict theory and numerous micro-sociologies developed as a number of sociologists lost interest in functionalism. During the 1980s a new functionalism, or neo-functionalism, began to emerge. Neo-functionalists mounted a determined effort to revive functionalism through a new interpretation of the work of Talcott Parsons. In particular, Alexander,[16] Colomy,[17] Sciulli,[18] and Sciulli and Gerstein[19] all sought to overcome the criticism that Parsons could not explain social change. Alexander, for example, modified Parsons' approach to incorporate economic rewards and political coercion, something to which Parsons paid little attention.[20] Thus, neo-functionalists have stressed the efforts of concrete groups in effecting social change through the use of power, conflict, and economic forces.

Functionalism and Humanist Sociology

Even if the neo-functionalists have adequately answered the criticism that functionalism does not explain social change, other criticisms render functionalism as having little value for humanist sociology. One of these criticisms is that functionalism is based on a metaphysical premise: that society has not only an existence of its own but also a mind or will of its own. Functionalists also assume that a society is goal directed and that these goals are devised by the society itself. Still another criticism is that the stability, unity, and harmony of societies tend to be exaggerated, and therefore existing social patterns and organizations are necessary and good. This produces a conservative bias—a fear of change—among functionalists. This can be seen readily in the list of functional imperatives, which typically include such items as the legitimacy of authority, social control, and the socialization of young people into traditional roles, that functionalists consider absolutely necessary to the maintenance of societies.

The most telling criticism of functionalism, from the standpoint of humanist sociology, is that functionalists do not seem to be aware of human freedom.[21] Parsons, Merton, and other functionalists are so conscious of stability that they assume individuals internalize and accept the rules governing social conduct in a society. Conformity is accepted as a given. Although Parsons talks about voluntarism, or choice, on the part of individuals as they strive for ends, the question of *whose* ends they are striving for is never asked. Alvin Gouldner states:

> Parsons never asks whether men are striving to achieve goals that they themselves have rationally inspected and selected, or whether theirs is the striving of tools, energetically seeking ends that others have programmed them to pursue. And Parsons never asks, under what social conditions can men select their own goals and under what condition will they blindly seek goals set for them by others? Parsons never sees that there is a profound difference between the failure to achieve one's own goals and the failure to achieve goals that others have imposed upon us. He fails to see that the ultimate alienation is not that we fail in what we seek, but that we seek what is not ours. The ultimate alienation is that we live our lives as tools and that we do not live for ourselves.[22]

Functionalism in general, and Parsonian functionalism in particular, condemns human freedom to irrelevance and so is of little use to humanist sociology.

Let us now look at conflict theory, another important contemporary sociological perspective, in order to see if it holds more promise for humanist sociology than does functionalism.

Conflict Theory

Given the conservatism of functionalism, in particular its emphasis upon stability and order in human society, a number of sociologists began to search for another perspective when the reality of the 1960s pointed up functionalism's inability to explain the abrupt changes then taking place. Conflict theory, which explains human behavior in terms of self-interest and the perpetuation of the social order by the organized coercion of various groups over others, became a rallying ground for sociologists who either had lost faith in functionalism or had really never accepted it. Drawing upon the works of Karl Marx (1818–1883) and Max Weber (1864–1920), contemporary conflict theorists sought to show that conflict theory is not simply a theory of conflict and disorder but "is also a theory of power, of social organization through the use of power, and of stability, legitimacy and social order."[23] Two major variants of contemporary conflict theory have emerged, one owing its origins to Marx, the other to Weber.

Karl Marx

Karl Marx divided society into two basic classes or strata, those who owned the means of economic production (the *bourgeoise*) and those who worked for the owners (the *proletariat*). Marxist analysis is based on the conflict between these two classes. Along with this emphasis upon class conflict, Marx contributed two other fundamental notions to the sociological perspective in general and conflict theory in particular. The first is his insistence that human societies be studied as wholes or totalities in which social groups, institutions, beliefs, doctrines, and so forth are interrelated. The second is his belief that societies are inherently alterable systems in which change is produced primarily by rational contradictions and conflicts.[24] These contributions are central to the conflict perspective.

Marx and his friend and collaborator Freidrich Engels provided an important foundation for later versions of conflict theory. However, because Marx based his views solely on experiences deriving from a capitalist economy, his analysis is somewhat limited. For example, to Marx, capitalism is essentially the product of two interrelated and antagonistic groups: the bourgeoise and the proletariat. Eventually, the antagonism between the two groups becomes so intense that a revolution occurs, resulting in the overthrow of the owning class by the working class. A communist society is then at hand, with the interests of all being reconciled. The problems with such a scenario are that capitalist societies have proven to be quite capable of modifying conflict. Communist societies that have come into existence since Marx wrote have not arisen through the overthrow of the bourgeoise by the proletariat. They instead have arisen through peasant revolutions in agrarian societies. And communist societies have imposed totalitarian governments upon the people (something that Marx abhorred) without achieving efficient levels of production or a just system of distribution.[25]

That such communist societies as East Germany, Poland, Czechoslovakia, Rumania, and the Soviet Union would remove their communist leaders in the late 1980s and early 1990s was also never envisioned by Marx, who held that communism would usher in "the end of history" in the form of the end of class conflicts.

Finally, Marx's horror of capitalism gets in the way of his analysis, pushing him into a deterministic view of human behavior. Marx began with the initial assumption that human nature was basically creative and cooperative, but he was faced with the reality that capitalist society was in conflict. How then did cooperative individuals come to engage in a competitive struggle with their fellows? Marx's answer was to pose the idea of an externally determining system of relationships among human beings. Capitalism's structures determined human behavior, and individuals were not allowed free choice. This is the contradiction in Marx, a dilemma which he could not solve.[26]

This should not, however, negate Marx's fundamental contribution to conflict theory. By emphasizing the totality, the interrelationship of human societies, and the widespread nature of conflict, contradictions, and change, Marx provided a valuable strategy point for conflict theory, one upon which Max Weber was to build.

Max Weber

A major concern of Max Weber, and what makes his work so important to conflict theory, was how power led to social order and integration. Power must always be considered if society is to be understood, for both conflict and social order are differentially derived from social power. The distribution of power, that is, legitimate power or authority, constitutes a continuing problem for society. Because power based solely on coercion is ineffective, society cannot remain stable unless the people see their leaders as legitimate. Thus, legitimate power is synonymous with authority.

Weber's views on legitimacy differed markedly from Marx's.

Whereas Marx constantly emphasized the increasing tendency on the part of the lower classes to view the government as illegitimate, in Weber's formulation, the government constantly seeks to enhance its legitimacy, losing it only when a powerful and charismatic leader capable of leading the people in a revolution comes upon the scene—a rare historical phenomenon. Therefore, the power structure is able to convince or coerce members of the society to accept the social order as legitimate even though it favors the interests of the more powerful.

Weber's contribution to conflict theory is a description of how power is legitimized and stabilized in society. Order rather than conflict does prevail in society as the functionalists maintain, but this order is achieved through the use of power to suppress or co-opt the underlying conflict.

Marx and Weber form the classical basis of conflict theory. However, it was contemporary theorists Ralf Dahrendorf (1929–) and Randall Collins (1941–) who, building upon the thought of these two giants, ushered in the interest in conflict theory in the 1960s and 1970s.

Ralf Dahrendorf

Ralf Dahrendorf's *Class and Class Conflict in Industrial Society*[27] can be said to have regenerated an interest in conflict theory unseen during the heyday of the functionalist-dominated post–World War II era. Starting with the insights of Max Weber, Dahrendorf offered a sympathetic review and reformulation of Marx's theoretical insights. In particular, Dahrendorf sought to demonstrate that Marx's conception of class, although perhaps viable in the nineteenth century, no longer held true. While Marx believed that one's class position resulted from one's relationship to the means of production, Dahrendorf believes that classes "are based on the differences in legitimate power associated with certain positions . . . or the structure of social roles with respect to these authority expectations."[28] The essential determinant of one's class position, then, is not economic ownership of property but power and legitimate authority. Although it is possible that in many individual cases wealth precedes authority, Dahrendorf argues that in the majority of cases an individual's position in a structure of authority precedes his or her ownership of capital. Where there is property, there is authority, but not every form of authority, in turn, implies property.[29] Prime examples would be General Colin Powell, who when he was head of the Joint Chiefs of Staff had little personal wealth but enormous power, or President Bill Clinton, who came to the White House with almost no personal wealth.

Priority, in Dahrendorf's scheme, is given to the concepts of power and authority, which then are used to develop a theory of conflict. For

Dahrendorf, the basic fact of social life is that individuals who live together have different interests that eventually come into conflict. Some gain greater power over others and then use this power to pursue and solidify their own interests. It is social power, in particular control over rewards and punishment, that organizes society and institutionalizes the basic conflict over interests present in any society.[30] Differential authority positions are found in all institutions, with the occupants of the dominant positions imposing their interests and values on those below them. It then follows that there always will be resistance to the exercise of power, but because the powerful are the stronger group, society is held together by the exercise of their strength through the coercion of those they dominate.[31]

Randall Collins

Randall Collins attempts to present conflict theory as a systematic whole.[32] Like Dahrendorf, he builds upon the works of Marx and Weber. Wealth, power and prestige are pursued in all societies, and everyone intensely dislikes being ordered around. Therefore, people will do their best to avoid this situation.[33]

These are basic facts of life—social conflict will always exist. "Since power and prestige are inherently scarce commodities, and wealth is often contingent upon them, the ambition of even a small proportion of persons for more than equal shares of [the] goods sets up an implicit counter-struggle on the part of others to avoid subjection and disesteem."[34] Given Collins' emphasis on coercion in social life, what are crucial for understanding social behavior are the (1) degree to which people are in positions to control others and (2) how this is related to the accumulation of wealth, power, and status.

Collins differs somewhat from Dahrendorf and most conflict theorists in the attention he gives to the individual. He is one of a few conflict theorists who begins with individuals and how they construct their world. For Collins, individuals live in self-constructed worlds where other people have the power to control them. This results in conflict.

Recent Developments in Conflict Theory

During the 1970s and 1980s, some neo-Marxists moved away from Marx's concern with an emancipatory breakthrough based on the emergence of the working class.[35] Two of the more influential neo-Marxist theories are the critical theory of Jurgen Habermas[36] and the world system approach of Immanual Wallerstein.[37] During this same time period, some feminist sociologists also began to develop feminist theory in the Marxist framework, while others developed an altogether new framework.

Jurgen Habermas and Critical Theory

Habermas sets out to refine Marx's thought, in particular the theory of historical materialism. He believes that human cognition and consciousness have evolved over centuries. In this view, human mental development is more important than materialism, because ideas set the historical framework within which economic and historical development take place. Evolution is heading toward an "ideal speech situation" in which everything can be fully discussed and agreed upon without coercion.[38] Freudian psychoanalysis contributes to the mental development necessary to achieve the self-insight for the ideal speech situation to occur.[39] For Habermas, the evolution of speech, language, and self-understanding emancipates individuals. The end product for Habermas is a rational society, where ideas are openly presented and defended against criticism.[40] The weight of argument, not force or power, determines what is considered to be valid or true.

Immanual Wallerstein

Wallerstein also takes a different approach from most Marxists and Weberian conflict theorists. Instead of looking at workers or classes, he concentrates on what he calls the *world-system*. According to Wallerstein, there have been only two examples of a world-system: the world empire of ancient Rome and the modern capitalist world economy.[41]

Wallerstein's theory rests on the relationship between the *core*, the *semi-periphery*, and the *periphery*. The core consists of social, political, and economic transactions among the world's dominant organizations, including the most developed nations, as well as multinational corporations and international banking interests. The semi-periphery is composed of regional individuals and groups who serve the interests of the core by managing the accumulation of raw materials and services for the periphery. The periphery consists of those parts of the world from which resources are extracted without receiving enough in exchange to allow for freedom and development.

Feminist Sociological Theory

The rise of the women's movement and an increase in the number of female sociologists has been accompanied by the creation and development of feminist social theory.[42] Taking a *sociology of knowledge perspective* (the view that ideas are shaped by social conditions) and incorporating such other disciplines as psychology, economics, history, and political science, feminist sociologists argue that male sociologists have failed to see the world from the female point of view. Feminists, therefore, begin with the vantage point of women in the social world. Feminist social theory is highly critical and value laden, arguing for a better world by asking: How can the world be changed and improved to be more just and humane for both women and men?[43]

Starting from the fundamental question, What about women? feminist theorists take gender as a central focus of their subject matter and question previous sociological theorizing for neglecting women. Dorothy Smith, for example, argues that male dominance through the control of language has produced a male sociology and calls into question whether feminist sociologists can practice feminist sociology using a male-dominated language.[44] Most feminists tend to agree with her interpretation.[45]

Calling for change, feminist sociologists present a *women-centered perception* of the world. Feminist sociologists see gender (the expectations attached to males and females) relations as problematic and in need of change.[46] The problems of gender differentiation and oppression are the consequence of a direct power relationship between men and women.[47] Patriarchy, the dominance of men, is seen as sustaining the basic power structure that suppresses women.

Conflict Theory and Humanist Sociology

Although conflict theory holds more promise for humanist sociology than does functionalism, there are some basic inconsistencies and problems inherent in this approach. In particular, the answer to the question of how much the individual is determined by social structures shows little consensus among conflict theorists. Although his early works on alienation are usually considered to have had a humanistic bent, in his later works Marx embraces a determinism in his attempt to forge a science of historical materialism. Dahrendorf is perhaps the most deterministic of the conflict theorists, seeing humans as essentially plastic beings. Individuals adapt to society because they are coerced, and even those who do the coercing do so with little choice in the matter. Dahrendorf specifically posits an individual who is born free but becomes the victim of structural constraint. The following is from Dahrendorf's essay "Homo Sociological."

> To become a part of society and a subject of sociological analysis, man must be socialized, chained to the fact of society and made its creature. By observation, imitation, indoctrination, and conscious learning, he must grow into the forms that society holds in readiness for him as an incumbent of positions. . . . For society and sociology, socialization invariably means depersonalization, the yielding up of man's absolute individuality and liberty to the constraint and generality of social rules.[48]

Even Collins, who starts from the individual and whose later works are an attempt to link the individual to the social structure, by opting for a science of sociology, comes down on the side of an individual who is more determined than determiner. Habermas, too, even though he calls for emancipation, by relying upon the thought of Sigmund Freud (who, as will be shown in chapter 3 is a determinist), in the end opts for a deterministic view of the human being. Feminists, alone, seem to posit a free individual as they analyze how women can overcome the gender restrictions imposed upon them. Some of their views will be incorporated into the humanist sociology that permeates this work.

In short, conflict theory presents a view of human behavior as shaped by powerful groups or classes. Those at the top exert control over those on the bottom. Although this insight is much more realistic than the functionalist view, which posits a smooth-running system based on consensus, conflict theory still does not go far enough—it does not consider the possibility that individuals can choose to overcome the constraints imposed upon them. Furthermore, most conflict theorists lack a valid theory of personality formation, a view which envisions human beings as both shaping and being shaped by the world.

Such a basis of personality theory, however, does exist in the form of symbolic interactionism. By combining the insights of the symbolic interactionists and the conflict theorists, a formulation of humanist sociology can be derived that realistically portrays the contemporary person as one who "need not be deformed into a mere tool of impersonal bureaucracies or a rash of power-seekers."[49]

Symbolic Interactions

Symbolic interactionism owes its principal origins to George Herbert Mead (1863–1931).

George Herbert Mead

The basic premise of Mead's social psychology is that the mind cannot develop without some form of social process or interaction between individuals.[50] The mind is not a physical substance with a specific location in the brain; rather it is something that functions in relation to what Mead called *significant symbols*. Significant symbols can be words, gestures, even grunts and groans. What makes something significant is consensus of meaning. If, at a party, a young man or a young woman smiles in a friendly manner at a member of the opposite sex whom he or she did not know and the smile is returned, this could be considered an invitation to engage in a conversation. If the two parties then begin to converse in a friendly fashion, then the smile was a significant symbol for both of them. If, however, one of the persons involved refuses to engage in a friendly conversation, then the smile is not a significant symbol, because it meant one thing to one and something else to the other.

Significant symbols arise only in some kind of social process. Individuals try to guess the intentions of others and then respond according to what they believe the others will do. Cooperative behavior among people is possible only if they can understand each other's actions and then guide their behavior accordingly. Human society rests upon this consensus, which is possible because of significant symbols.

To return to the party example, if the person who first smiled continues to smile at other people, then refuses to converse with them, no cooperative behavior would be possible. Unless the person did not want to make new acquaintances, he or she would have to change his or her behavior so that others could understand his or her intentions. This change could be anything from winking to bringing over a drink, as long as both parties understood what was occurring.

For Mead, we human beings acquire a mind in the process of developing the capacity to create significant symbols. The actions of others are completed in our heads. A man sees someone smiling at him, thinks to himself that the other person wants to make his acquaintance, and he acts accordingly. He has put himself in the other's place, or to use Mead's term, he role-plays.

Individuals modify and control their own behavior in terms of this role playing. Thinking occurs as individuals engage in conversations with themselves and imagine how others would respond to them. Language is of primary importance, because it alone stimulates speaker and hearer in the same manner and speeds the process of developing significant symbols. This mechanism, according to Mead, is found only in human society. Vocal gestures provide the method by which social organization in society becomes possible. The young man or the young woman smiles at members of the opposite sex at a party in order to make their acquaintance.

Language and the mechanisms of communication together with role playing result in a social conception of mind. Mead formulated a theory which produced a shift in the then basic assumptions underlying the nature and function of language. With his friend and colleague, the famous American philosopher John Dewey, he pioneered a new approach, in which the nature of language is not the expression of prior thought but is essentially communication—the establishment of cooperation in social activity—in which the self and the other are modified and regulated by common action.

The self becomes social through language, which enables the individual to role-play. At first, children play at different roles, imitating the actions of the people they see and moving in and out of their roles. Later, they begin to play at organized games. Here the child must be ready to take the attitude of everyone else in the game and must be aware that these different roles have a definite relationship to each other. For example, when the child participates "in a baseball game, he must have the responses of each position involved in his own position. He must know what everyone else is going to do in order to carry out his own play. He has to take all of these roles."[51]

The crystallization of all these attitudes and responses is what Mead termed the *generalized other*. This he defined as: "The organized community or social group which gives to the individual his unity of self. . . . The attitude of the generalized other is the attitude of the whole group."[52]

Consciousness and will arise from problems. Individuals ascertain the intentions of others and then respond on the basis of their interpretations. If there were no interactions with others, there would be no development of the mind. Individuals possess the ability to modify their own behavior; they are subjects who construct their acts rather than simply respond in predetermined ways. Human beings are capable of reflexive behavior, that is, turning back and thinking about their experience. The individual, according to Mead, is not a passive agent who merely responds to external constraints but someone who chooses between alternatives. As Meltzer, Petras, and Reynolds phrase it:

> For Mead, the role played by the individual was one of interpreting data furnished him or her in the social situation. Thus, while blocked impulses were satisfied according to the individual's own desires, choices of potential solutions were bounded by the given facts of his or her presence in the larger network of society. The presence in two systems made men and women both determined and determiners.[53]

These insights of Mead's have been refined in the work of the most important contemporary symbolic interactionist, Herbert Blumer (1900–1986).

Herbert Blumer

Herbert Blumer, who coined the term "symbolic interactionism," sees symbolic interactionism as referring to "the peculiar and distinctive character of interaction as it takes place between human beings."[54] In this approach human beings do not simply react to each other but interpret and define each other's actions. For Blumer, symbolic interactionism has three basic premises:

(1) human beings act toward things on the basis of the meanings that the things have for them; (2) these meanings are a product of social interaction in human society; and (3) these meanings are modified and handled through an interpretive process that is used by each person in dealing with the things they encounter.[55]

Human beings construct reality through symbolic interaction with others. In the process, individuals must get "inside the head" of the other person if they are to understand what the person is doing.

To Blumer, humans have a self that they can make the object of their own actions; that is, they can act toward themselves just as they can act toward others. This ability enables individuals to make indications to themselves, a process in which they note things, assess them, give them meaning, then decide to act on the basis of this meaning. These *self-indications* constitute what is referred to by Blumer as consciousness. Individuals are conscious of things ranging from the ticking of a clock to abstract reasoning. All of these are self-indications, and the conscious life of an individual is a steady flow of such self-indications.[56]

The self is a dynamic process of interaction. It is a reflexive operation in which individuals take note of the world around them and ascertain its importance for their eventual actions. Individuals rehearse their behavior, summon plans of action, assess them, change them, and form new ones, all the while indicating to themselves what their course of action will be. This tentative, exploratory process provides for the possibility of novel behavior.[57] By viewing behavior as the interplay between the spontaneous and the socially determined aspects of self, Blumer envisioned human behavior as innovative, unpredictable, and indeterminate.

The possibilities for a humanist sociology seem readily apparent in Blumer's symbolic interactionism. However, substantive criticisms of this approach have been tendered, and they must be examined in order to see just how much of symbolic interactionism is viable for a humanist sociology. This will be done after we look at recent developments in the symbolic interactionist perspective.

Recent Developments in Symbolic Interactionism

A major criticism of symbolic interactionism is its neglect of large-scale social structures. Sheldon Stryker, in *Symbolic Interactionism* (1980), has sought to remedy this lack of attention. Stryker states: "A satisfactory theoretical framework must bridge social structure and persons, must be able to move back from the level of the person to that of large-scale social structure and back again."[58] Stryker accords central importance to larger social structures, conceiving of them in terms of organized patterns of behavior. Social structure is seen by him as the "framework" within which people interact. People are constrained by social structures, but they do not simply take roles. Instead, they engage in "role-making," which entails an active, creative orientation to their roles. Social structures, however, limit the amount of role-making that can be accomplished. Some social structures permit more than others. It is this possibility of role-making that produces changes in expectations, and the cumulative effect of these changes produces changes in larger social structures.

Robert Perinbanayagam, another symbolic interactionist, introduced the concept of "programs or plans to be realized in time and moves to be initiated for a later moment."[59] Programs are not mental constructs but associations like families, schools, and bureaucracies. Programs are linked to the larger society through the number of people involved and the availability to, or the exclusion of, various groups. Thus social class, ethnic status, age, and gender all allow or curtail access to certain programs.

Symbolic Interactionism and Humanist Sociology

A number of criticisms have been raised in the literature concerning symbolic interactionism. Foremost among these is the one stated earlier as to whether symbolic interactionism either ignores or has a faulty conception of social organization and social structure. This is a sound criticism, one that neither Stryker, Perinbanayagam, nor any other symbolic interactionist adequately answers. Symbolic interactionism could use a more defined structural perspective as presented, for example, in conflict theory. A synthesis of symbolic interactionism and conflict theory could provide the basis of a humanist sociology.

Another criticism concerns symbolic interactionism's methodological problems. Its concepts are difficult to operationalize, it generates too few testable hypotheses, and it has failed to formulate specific methodological procedures. This criticism is the prevailing view of mainstream sociology. Humanist sociology rejects the notion that the scientific method is the *only* method for achieving knowledge in the social sciences. In chapter 2, other techniques and methods which do not rely upon operational definitions, testable hypotheses, and advanced statistical techniques are discussed. For now, it is enough to say that such alternate methods of data collection as unobstrusive measures, participant observation, and sympathetic understanding, among several, do exist and offer viable alternatives to positivism.

A third criticism of symbolic interactionism is that it plays down the unconscious and generally ignores such psychological dimensions as needs, motives, intentions, and aspirations, treating them simply as mere derivations and/or expressions of socially defined categories. This is an exceedingly well-taken criticism, which should be applied not only to symbolic interactionism but also to sociology in general. Sociologists have refused to investigate the psychological dimension of behavior, thereby overlooking an important factor in the explanation of human action. In chapter 3, some of the psychological theories of behavior are discussed and incorporated into a humanist sociology in order to make it broader than traditional sociological approaches.

A fourth criticism relates to the key concept of symbolic interactionism—the self—being vaguely defined and therefore becoming the object of much confusion. This is another good criticism. Symbolic interactionism has failed to refine the insights of George Herbert Mead. This is discussed further in chapter 3 where Mead's social behaviorism is combined with the thoughts of certain contemporary theorists to form a basis upon which to build a humanist theory of self.

A fifth criticism of symbolic interactionism is that it is ahistorical. Specific phenomena under study are rarely linked to their historical origins and development. This, too, is a well-taken criticism. A humanist sociology should take into consideration the historical dimensions of the social phenomena or problems under scrutiny. Such an historical dimension is offered in chapters 6 and 7, in which the institutions of society—political, economic, religious, family, educational, and the mass media—and their relationship to the contemporary American social organization are discussed.

A final criticism of symbolic interactionism is that it has a limited view of the nature of social power. Because social power is connected to social organization and social structure, a synthesis of symbolic interactionism and conflict theory offers the best starting point for a telling humanist sociology. But before the presentation of this synthesis, one more thinker must be presented—C. Wright Mills (1916–1962). Mills was a sociologist who perhaps more than anyone else in this century embodied the qualities of a humanist sociologist.[60]

Wright Mills as an Exemplar of Humanist Sociology

Virtually alone, C. Wright Mills rallied against the traditional, conservative, fuctionalist sociology that characterized the 1950s. He attempted to fashion a critical, humanist sociology—which would help liberate individuals—through a synthesis of George Herbert Mead's social psychology (symbolic interactionism) and a Marxian-Weberian version of social structure (conflict theory).

Mills' most precise articulation of humanist sociology is found in *Character and Social Structure* (1953),[61] written with Hans Gerth, and in *The Sociological Imagination* (1959),[62] which is essentially a reformulation of the framework worked out in *Character and Social Structure*. The major difference between the two works is that *The Sociological Imagination* is more action-oriented, has a greater emphasis on the historical location of particular social structures, and has a lesser emphasis upon personality formation. Mills had worked out a viable system of personality formation in *Character and Social Structure*. In his subsequent works, he concentrated more on objective factors, what he referred to as "the main drift" of those historical and structural forces that were often impersonal and unrecognized by those who suffered their impact.[63]

Mills rejected both the assumption that humans were passive recipients of external sense data and the belief that the mind contained innate ideas fixed at birth. He began with the view of the individual as a biological entity who also possesses a psychic structure. This psychic structure, or personality, is composed of integrated feelings, sensations, and impulses. These elements are part of our biological heritage, but their specific transformation into emotions, perceptions, and purposes can be understood by focusing upon the individual as a player of roles, as a "person," to use Mills' term.

Because human beings are social creatures, they must be analyzed in terms of their social actions. Thus, human behavior can be understood in terms of socially constructed motives rather than as responses to external stimuli. Differences found among individuals can be attributed to the constitution of their organisms, to the specific roles they play, and to the novel and idiosyncratic integration of their personalities within their psychic structure.

In order to link the individual to a conception of social structure or society, Mills incorporated Max Weber's notion of "social relationship." A social relationship exists, according to Weber, when "the behavior of a plurality of actors insofar as, in its meaningful content, the action of each takes account of that of the others and is oriented in these terms."[64] The Weberian social structure is one of probable patterns of behavior. Individuals expect that others will act in a certain way, and they adjust their own behavior accordingly. As long as there is a probability that one's behavior will meet with the expected reactions and viceversa, a social structure exists, and order comes into being. Weber thus shows how social conduct gives rise to other actions that become the basis for political, economic, religious, and other organizations.[65]

Mills used the concept of role to bridge the gap between the individual and the Weberian notion of social structure. Roles are by definition interpersonal or geared to the expectations of others.[66] These others are also playing roles, and the mutual expectations set up patterns of social conduct. The individual's psychological functions are thus shaped by specific configurations of roles incorporated from the society.

The most important aspect of personality is the individual's concept of self, or the image one has of what kind of person one is. Thus, the roles one plays and the image one holds of one's self are firmly entrenched in the social context. The psychological characteristics of an individual are linked to the controls of a society by the relation of role to institution. An institution is defined as "an organization of roles, . . . one or more of which is understood to serve the maintenance of the total set of roles."[67] Institutions, in turn, make up institutional orders that serve functions. Mills lists the major institutional orders as political, economic, military, religious, and kinship (family). Thus, a social structure is comprised of the major institutional orders and their relationship to each other.

Mills offers a sociology that is liberating while still possessing an adequate conception of social structure—one that does not sacrifice the volitional, active nature of human beings. Mills points out a path toward freedom. Because of his early philosophical training in pragmatism (the specifically American philosophical system that emphasized the finding of truth in what worked), Mills never gave up the notion of the autonomous individual who could use reason to gain and secure freedom. Like Marx, Mills believed that humans were alienated, given the society they lived in, but where Marx saw alienation as the result of the irrationality of production, Mills saw it coming from a perception of and adaptation to a society that resulted from blind drift. In order to be free the individual had to make the connection between "private troubles" and "public issues." He or she had to be aware that structural problems were the key to his or her malaise. Only by seeing the interconnection of biography and history could individuals begin to gauge the limits of their potential. This was the fundamental message of The Sociological Imagination. The social scientist must look at the structure of society as a whole and at the ways in which institutions that comprise it shape the character of individuals. But more than this, Mills argued, the social scientist must "study the structural limits of human decision in an attempt to find points of effective intervention, in order to know what can and what must be structurally changed if the role of explicit decision in history-making is to be enlarged."[68] The possessor of the sociological imagination should study historical structures in order to find more and more ways which can insure the freedom of individuals.

Mills' humanist sociology enables the individual to transcend the realm of private troubles, to see that structural problems are at the root of alienation. Reason could lead to freedom when and if the individual became aware that "rationally organized arrangements . . . often . . . are means of manipulation."[69] It is for the humanist sociologist to point out how people can be free in their society—to try to insure freedom by intervening to make the structure of society less repressive.

Mills offers a picture of human beings as potentially free but constrained by power relations. In this view some people (those with power) are freer than others and are therefore responsible for their actions. History is made behind people's backs but not behind everyone's back. There are varying degrees of freedom. What Mills offers is a doctrine of moral responsibility in the face of societal constraint, in essence, a study of how freedom is being eroded.

This vision of humanist sociology is also apparent in Mills' other major works. In *White Collar* (1956), for example, it underlies Mills' analysis of the massive changes that had taken place in the class structure of the United States since the nineteenth century. *White Collar* is a social-psychological study of how the bureaucratization of the economic institutional order affects the social biographies of those individuals who act out social roles within this order.[70]

The portrait painted in *White Collar* is a bleak one. White-collar workers are not free within the economic institutions, and this lack of freedom becomes more and more dominant in their character structure as they habitually submit to the will of others. The individual was seen by Mills as trapped by the drift of modern capitalism. The "new middle class" had arisen out of an occupational shift from independent entrepreneurs to white-collar workers. The general result was the formation of a powerless social type cast adrift in a society that was fast becoming a mass society. The new middle class had defaulted in the struggle for power. The locus of power resided in the hierarchies of large-scale institutions. The white-collar worker had inherited a world of alienation from power, from work, and from self. The middle class was so divorced from power that alienation, anxiety, and insecurity had become general psychological traits.

In *White Collar*, Mills put together a cohesive synthesis of Meadian social-psychology and Weberian sociology. The Weberian notion of the bureaucratization of society is imposed upon the personality structure of those who live and work within the large-scale institutions. The character of individual men and women as shaped by their particular social structure forms the content of *White Collar*.

The Power Elite (1956), like *White Collar*, was based upon a conception of the institutional integration of society. Mills' analysis again was social-psychological, one in which he sought to relate the psychology of the individual to social structure via the intermediary of stratification. American society was ruled by a ruling stratum, or a power elite. The United States was integrated through the interlocking domination of the political, economic, and military orders. Power was located at the top of these orders. The power elite were thus similar types and had similar personality characteristics, since they had been shaped by the same institutional configurations.[71]

Mills followed *The Power Elite* with *The Causes of World War III* (1958), in which he tried to show that the power elite was leading the United States into a total and absurd war.[72] He was by then convinced of his own general thesis that the United States no longer consisted of a self-contained economy, an autonomous political order, and a subservient military. The power elite had to be understood in connection with the development of a permanent war establishment, alongside a privately incorporated economy and inside a virtual political vacuum. This was not the realization of any plot but simply the result of structural trends. Power and political responsibility resided with the upper stratum. Their consciousness, their views of reality, were shaped by their position in society. And they were leading the United States toward World War III.

In the three years between the publication of *The Sociological Imagination* (1959) and his death in 1962, Mills published three books. He edited a collection of essays of the founding fathers of sociology, entitled *Images of Man* (1960), and followed this with *Listen, Yankee* (1960) and *The Marxists* (1962).[73] *Images of Man* represents Mills' acknowledgment of his own intellectual debts and is important for the understanding of the antecedents of his humanist sociology. *Listen, Yankee*, on the other hand, is a disappointment for two reasons. The first is that Mills himself referred to it as a "pamphlet," and the work, which he wrote in six weeks after visiting Cuba, never rises above this level. Second, Mills portrayed the revolution in Cuba as an example of the intellectual being the vanguard for change and Castro as an intellectual hero; but he began to believe that Castro had deceived him and was debating with himself just before his death whether to denounce *Listen, Yankee*.[74]

The Marxists is the most important of his later books, for here Mills offered a perceptive critique of both liberalism and Marxism. Liberalism, in his view, implied an inability to see the whole structure of society; it had no grasp of history. To Mills, liberalism was the ideology of the entrepreneurial middle class. Marxism, on the other hand, did have an overall view, but it was an inadequate one.

Though he considered Marx *the* political thinker of the nineteenth century, *that* thinker whom social scientists had to be familiar with to be even considered social scientists, this did not preclude his offering a trenchant criticism of Marx. Mills considered Marx a determinist and stated this explicitly in *The Marxists*.[75] Given his emphasis on freedom, any form of determinism was anathema to Mills. His final conclusion regarding Marxism was that the general trend of history had rendered much of Marxist theory inadequate.

Mills, therefore, is an exemplar of humanist sociology. He tried to achieve the promise of the Enlightenment—the fusion of morals with social science—and tried to show the "direction" such a social science should take.

Mills' humanist sociology advocates a moral responsibility to create a just and humane society. The main purpose in amassing a body of knowledge is to serve human needs; knowledge ought to be useful. By adopting this position, Mills extends the analysis of what *is* to the analysis of what *ought* to be. Knowledge is the mechanism to provide a humane society.

Responsibility as a moral standard is an integral part of Mills' humanist sociology. A fundamental quality of human beings is their potentiality for ethical autonomy. People not only *are* but *ought* to be in charge of their own destiny within the limits permitted by their social structures.

Mills' humanist sociology is predicated on the belief that individuals must engage in those forms of social action that enable them to be active participants and not passive spectators in their own destinies. Mills presented the beginnings of a humanist sociology which takes account of the volitional, active nature of the person. To be free, people should be aware of the constraints that society uses to thwart their potential. Sociologists should not be detached observers, scientists who claim to stand outside what they study, but men and women who actively participate in the attempt to liberate other human beings.

From this perspective we can begin to fashion a humanist sociology. But it must be borne in mind that Mills' sociology is just a beginning and not a finished product by any means. In particular, his views on the development of the self and roles and the number of institutions and their relationship to each other must be clarified and refined. An attempt will be made to do this.

Summary

This chapter introduced humanist sociology, a sociology based on the principles of the insurance and maximization of human freedom. The major perspectives in contemporary sociology, functionalism, conflict theory, and symbolic interactionism were reviewed in light of this humanist sociology. Functionalism was seen as condemning human freedom to irrelevance. Conflict theory was found to offer a view of humans as essentially plastic beings who adapt to society because of coercion. However, given insights into society as being held together by the powerful coercing the less powerful, conflict theory was seen as holding some promise for humanist sociology. Symbolic interactionism offers a dynamic view of potentially free human beings but lacks the emphasis on social structures provided by conflict theory. By combining conflict theory with symbolic interactionism, a humanist sociology is possible. The work of C. Wright Mills, who attempted to do just this, was presented as an important starting point for this synthesis of conflict theory and symbolic interactionism, or the humanist sociology that is presented in this text.

Notes

1. Ernest Becker, *The Lost Science of Man* (New York: Braziller, 1971), p. 121.

2. This is a paraphrase of French sociologist George Gurvitch's definition, which reads: "Sociology is the science of human freedom and all of the obstacles which this definition encounters and overcomes in part." Quoted in Paul Filmer, Michael Phillipson, David Silverman, and David Walsh, *New Directions in Sociological Theory* (Cambridge, MA: MIT Press, 1973), p. 124.

3. See, for example, Peter L. Berger, *Invitation to Sociology: A Humanistic Perspective* (Garden City, NY: Anchor Books, 1963); John F. Glass, "The Humanistic Challenge to Sociology," *Journal of Humanistic Psychology* 11(1971):170–83; and Alfred M. Lee, *Toward Humanist Sociology* Englewood Cliffs, NJ: Prentice-Hall, 1973. Each offers a different approach to humanist sociology.

4. See especially *The Positive Philosophy of Auguste Comte*, transl. and condensed by Harriet Martineau (London: J. Chapman, 1853).

5. Herbert Spencer, *The Study of Sociology* (New York: Appleton, 1891), p. 394.

6. Emile Durkheim, *The Division of Labor in Society*, George Simpson, trans. (Glencoe, IL: Free Press, 1956); originally published in 1893.

7. Emile Durkheim, *The Rules of Sociological Method* (Glencoe, IL: Free Press, 1950), p. 97; originally published in 1895.

8. Emile Durkheim, *Suicide* (Glencoe, IL: The Free Press, 1951); originally published in 1897.

9. Ibid.

10. Durkheim, *Rules*, p. 97.

11. Lewis A. Coser, *Masters of Sociological Thought*, 2d ed. (New York: Harcourt, Brace, Jovanovich, 1977), p. 142.

12. See especially Talcott Parsons, *The Social System* (New York: Free Press, 1951).

13. Robert K. Merton, *Social Theory and Social Structure* (New York: Free Press, 1963).

14. Ibid.

15. Jeffrey Alexander, "Toward Neo-Functionalism," *Sociological Theory* 3(1985):11–23; and "Neofunctionalism Today: Reconstructing a Theoretical Tradition," in George Ritzer, ed., *Frontiers of Social Theory* (New York: Columbia University Press, 1990).

16. Alexander, "Toward Neo-Functionalism."

17. Paul Colomy, "Recent Developments in the Functionalist Approach to Change," *Sociological Focus* 19(1986):139–58.

18. David Scuilli, "Voluntaristic Action as a Distinct Concept: Theoretical Foundations of Social Constitutionalism" *American Sociological Review* 51(1986):743–66.

19. David Scuilli and Dean Gerstein, "Social Theory and Talcott Parsons in the 1980s," *Annual Review of Sociology* 2(1985): 369–87.

20. Jeffrey Alexander, *Theoretical Logic in Sociology.* Vol. 4: *The Modern Reconstruction of Classical Thought: Talcott Parsons* (Berkeley: University of California Press, 1983).

21. Anthony Giddens, *The Constitution of Society* (New York: Columbia University Press, 1984); Alvin W. Gouldner, *The Coming Crisis of Western Sociology* (New York: Basic Books, 1970).

22. Gouldner, *Coming Crisis*, p. 193.

23. James Duke, *Power and Conflict in Social Life* (Provo, UT: Brigham Young University Press, 1976), p. 4.

24. Thomas B. Bottomore, "Marxist Sociology," in *The International Encyclopedia of the Social Sciences*, vol. 10, David L. Sills, ed. (New York: Macmillan, 1968), p. 46.

25. Joseph A. Scimecca and Arnold K. Sherman, *Sociology: Analysis and Application* (Dubuque, IA: Kendall/Hunt, 1992).

26. For an elaboration of this point, see Dick Atkinson, *Orthodox Consensus and Radical Alternative* (New York: Basic Books, 1972), p. 36.

27. Ralf Dahrendorf, *Class and Class Conflict in Industrial Society* (Stanford, CA: Stanford University Press, 1959).

28. Ibid., p. 149.

29. Ibid., p. 147.

30. Duke, *Power and Conflict*, p. 153.

31. Ralf Dahrendorf, "In Praise of Thrasymachus," in *Essays in the Theory of Society* (Stanford, CA: Stanford University Press, 1968), p. 134.

32. Randall Collins, *Conflict Sociology: Toward an Explanatory Science* (New York: Academic Press, 1975); *Theoretical Sociology* (San Diego, CA: Harcourt Brace Jovanovich, 1988); and "Conflict Theory and the Advance of Macro-Historical Sociology," in Ritzer, *Frontiers of Social Theory*, pp. 68–87.

33. Collins, *Conflict Sociology*.

34. Randall Collins, "Functional and Conflict Theories of Educational Stratification," *American Sociological Review* 36(1971):1009.

35. Robert J. Antonio, "The Problem of Normative Foundations in Emancipatory Theory: Evolutionary versus Pragmatic Perspectives," *American Journal of Sociology* 94(1989):721–48.

36. Jurgen Habermas, *Toward a Rational Society* (Boston, MA: Beacon Press, 1970); *Knowledge and Human Interests* (Boston, MA: Beacon Press, 1971); *Legitimation Crisis* (Boston, MA: Beacon Press, 1975); *Communication and the Evolution of Society* (Boston, MA: Beacon Press, 1979); *The Theory of Communicative Action*, vol. 1 (Boston, MA: Beacon Press, 1984); and *The Theory of Communicative Action*, vol. 2 (Boston, MA: Beacon Press, 1987).

37. Immanual Wallerstein, *The Modern-World System: Capitalist Agriculture and the Origins of the European World-Economy in the Sixteenth Century* (New York: Academic Press, 1974); *The Modern-World System II: Mercantilism and the Consolidation of the European World-Economy, 1600–1740* (New York:

Academic Press, 1980); and *The Modern-World System III: The Second Era of Great Expansion of the Capitalist Economy, 1730–1840* (San Diego, CA: Academic Press, 1989).

38. Habermas, *Knowledge and Human Interests*.

39. Habermas, *Legitimation Crisis*.

40. Habermas, *Communication and the Evolution of Society*.

41. Wallerstein, *The Modern-World System*.

42. Janet S. Chafetz, *Feminist Sociology* (Itasca, IL: Peacock, 1988); Patricia M. Lengermann and Jill Niebrugge-Brantley, "Feminist Sociological Theory: The Near-Future Prospects," in Ritzer, *Frontiers of Social Theory*, pp. 316–44; Patricia M. Lengermann and Ruth A. Wallace, *Gender in America: Social Control and Social Change* (Englewood Cliffs, NJ: Prentice Hall, 1985); and Dorothy E. Smith, *The Everyday World as Problematic: A Feminist Sociology* (Boston, MA: Northeastern University Press, 1987), and *The Conceptual Practices of Power: A Feminist Sociology of Knowledge* (Boston, MA: Northeastern University Press, 1991).

43. Smith, *The Everyday World as Problematic* and *The Conceptual Practices of Power*.

44. Ibid.

45. See Lengermann and Niebrugge-Brantley, "Feminist Sociological Theory."

46. Chafetz, *Feminist Sociology*.

47. Carol Gilligan, *In a Different Voice: Psychological Theory and Women's Development* (Cambridge, MA: Harvard University Press, 1982); Alison M. Jagger, *Feminist Politics and Human Nature* (Totowa, NJ: Rowman and Allenhand, 1983).

48. Ralf Dahrendorf, "Homo Sociologicus," in *Essays in the Theory of Society*, pp. 56–57.

49. Lee, *Toward Humanist Sociology*, p. ix.

50. George Herbert Mead, *George Herbert Mead on Social Psychology*, Anselm Strauss, ed. (Chicago, IL: University of Chicago Press, 1956).

51. Ibid., p. 215.

52. Ibid., p. 218.

53. Bernard N. Meltzer, John W. Petras, and Larry T. Reynolds, *Symbolic Interactionism: Genesis, Varieties and Criticism* (Boston, MA: Routledge & Kegan Paul, 1977), pp. 41–42.

54. Herbert Blumer, "Society and Symbolic Interactionism," in Arnold Rose, ed., *Human Behavior and Social Process*. Boston, MA: Houghton-Mifflin, 1962, p. 180.

55. Herbert Blumer, *Symbolic Interactionism* (Berkeley, CA: University of California Press, 1986).

56. Blumer, "Society and Symbolic Interactionism."

57. Meltzer, Petras, and Reynolds, *Symbolic Interactionism*, p. 54.

58. Sheldon Stryker, *Symbolic Interactionism* (Menlo Park, CA: Benjamin/Cummins, 1980), p. 53.

59. Robert S. Perinbanayagam, *Signifying Acts: Structure and Meaning in Everyday Life* (Carbondale: Southern Illinois University Press, 1985).

60. The interpretation of Mills in the following section is based on Joseph A. Scimecca, *The Sociological Theory of C. Wright Mills* (Port Washington, NY: Kennikat Press, 1977); and "The Philosophical Foundations of Humanist Sociology," in John Wilson, ed., *Current Perspectives in Sociological Theory*, vol. 9 (Greenwich, CT: JAI Press, 1989), pp. 223–38.

61. Hans Gerth and C. Wright Mills, *Character and Social Structure* (New York: Harbinger Books, 1953/1964).

62. C. Wright Mills, *The Sociological Imagination* (New York: Oxford University Press, 1959).

63. Scimecca, *The Sociological Theory of C. Wright Mills.*

64. Max Weber, *The Theory of Social and Economic Organization*, Talcott Parsons, ed. (New York: Oxford University Press, 1947), p. 118.

65. Julien Freund, *The Sociology of Max Weber* (New York: Pantheon Books, 1968).

66. Gerth and Mills, *Character and Social Structure*, pp. 10–22.

67. Ibid., p. 13.

68. Mills, *Sociological Imagination*, p. 174.

69. Ibid., p. 169.

70. C. Wright Mills, *White Collar* (New York: Oxford University Press, 1951).

71. C. Wright Mills, *The Power Elite* (New York: Oxford University Press, 1956).

72. C. Wright Mills, *The Causes of World War II* (New York: Simon and Schuster, 1958).

73. C. Wright Mills, ed., *Images of Man* (New York: George Braziller, 1960); *Listen, Yankee: The Revolution in Cuba* (New York: Ballantine Books, 1960); *The Marxists* (New York: Dell, 1962).

74. Harvey, Swados, "C. Wright Mills: A Personal Memoir," *Dissent* 10(1963):35–42.

75. Mills, *The Marxists.*

CHAPTER 2

Methods of Social Research

Sociologists, like all social scientists, are interested in explaining human behavior. In so doing, they use a variety of research methods to test their understanding. Sociologists, however, have split over which research procedure is best. Some argue that sociology can develop according to Auguste Comte's philosophy of *positivism*—the view that all phenomena are subject to invariable natural laws and principles like those found in the natural and physical sciences.[1] Others argue that human behavior is too complex and unpredictable to be studied in the same manner as are physical and natural objects.

Up until the 1960s and 1970s, most sociologists supported the former position and relied on positivistic research methods that presupposed a deterministic interpretation of social phenomena. To the positivist, individuals have little or no choice in their social behavior; they follow social laws just as physical elements follow physical laws. Positivists see little difference between the research techniques they use and those of their colleagues in the natural and physical sciences—their respective objects of study are deemed equivalent.

Beginning in the 1960s the dominant position of positivism in sociology and all social sciences became subject to a variety of criticisms. Philosophical movements such as hermeneutics, ordinary language philosophy, poststructuralism, and critical theory, to mention some of the more prominent, converged in rendering problematic the basic assumptions of positivism. All espoused the position that the social sciences involve an interpretive ordering of social reality.[2]

In response to this criticism, the strict positivism of Comte was replaced by a more sophisticated kind, best exemplified in

29

the work of Jonathan Turner. Turner offers five modifications of strict positivism.

1. Although sociologists, at times, may be able to state relations mathematically, deductions will rarely be in terms of a calculus of mathematics or some other system of formal logic. Sociologists should be content with laws that are stated precisely and with juxtapositions of the law with data in a way that seems reasonable and justified.
2. The criterion of prediction is unrealistic for any science that cannot test most of its laws in experimental situations or under circumstances where extraneous factors can be eliminated or, alternatively, known and measured.
3. Although falsification is important, it is not a lock-step procedure but is a more fluid process of negotiating and assessing plausibility.
4. Although positivism is concerned with assessing theory with data, assessment is not an inductive process. The goal of sociology is to generate theory first and then test its empirical merits.
5. The question of causality must not be ignored. Even though in some ultimate philosophical sense, cause cannot be known, nevertheless the sociologist must conceptualize forces in the universe as affecting each other in patterned and causally connected ways.[3]

Positivist sociologists, even the less strict ones, have defined understanding and exploration almost exclusively in quantifiable terms, holding that anything that cannot be verified through measurement is not worth knowing. In this view of the sociologist as scientist, there also is the image of the value-free investigator, one whose own biases and ideological leanings do not enter into the research endeavor. Knowledge and scientific inquiry are neutral.

Humanist sociologists reject the positivist view and hold that human beings are not subject to general laws, precise or imprecise. Humanist sociologists assume that there are basic differences between chemical elements and human beings that are not reducible to degrees of control over experimentation and external forces. Human beings are aware of what happens to them and can reflect on this and thereby alter the course of events (in other words, they possess freedom). To the humanist sociologist, positivism and its methods of research have limited use in assessing the reality of the social world. Quantification, the staple of positivism, has not led to the heightened understanding its practitioners seek.

Humanist sociologists attempt to understand human behavior by trying to see social reality from the perspective of the thinking individual. A basic premise of humanist sociology is that in order to explain social reality, the researcher must come to understand what the early twentieth-century sociologist W. I. Thomas called the individual's "definition of the situation,"[4] or an individual's perceptions and interpretations of the world and how this affects that person's behavior. Individuals attach meaning to the social world and their lives, and this is what differentiates human behavior from animal and physical behavior, sociology from biological and natural science. Because of this fundamental distinction, humanist sociologists advocate a method of analysis other than positivism, that of "qualitative methods."

Strauss and Corbin define qualitative methods as:

> any kind of research that produces findings not arrived at by means of statistical procedures or other means of quantification. It can refer to research about persons' lives, stories, behavior, but also about organizational functioning, social movements, or interactional relationships. Some of the data may be quantified as with census data but the analysis itself is a qualitative one.[5]

Here it must be noted that qualitative methods as used by the humanist sociologist are alternatives to and not replacements for quantitative methods. Humanist sociologists advocate *methodological pluralism*, the belief that there is no one right way of doing sociological research. It is what the sociologist seeks to describe and explain that should determine the research method used and not vice versa.[6] There are a variety of right methods, each offering part of the picture of reality being analyzed. Some phenomena lend themselves to one type of research method, others to another. Although there is no place within humanist sociology for a strict deterministic positivism, there is a place for quantitative methods, for they do inform us about the world. As will be seen in chapter 5, much of the understanding of social class and social inequality is based on studies which use the latest quantitative research techniques developed by positivist sociologists. However, positivism still is a limited view, and sociology has suffered because it has so often been considered the only accepted method of sociological investigation. Humanist sociologists should not make the mistake that positivist sociologists have made in assuming that there is just *one* sociological method. While humanist sociologists hold that qualitative methods lead to a greater understanding of human behavior, quantitative methods should not be dismissed out of hand. Therefore, each type of method will be reviewed with the idea that a fuller understanding of the limits and strengths of both perspectives will lead to a better explanation of the complexity that is social behavior.

Quantitative Methods of Sociological Research

Quantitatively oriented sociologists most often use questionnaires or surveys. The researcher constructs a series of questions, then asks a sample of people who are thought to have the same characteristics as the larger group of which they are a part what they think about certain things, how they have acted in the past, how they think they would act in the future, and so forth. From these answers, sociologists draw statistical inferences, which then lead to generalizations about the larger group. The social world, positivists assert, exists objectively; it is "out there," independent of the observer, and can be understood through the use of a well-constructed questionnaire. Regularities and patterns can be established if the questionnaire or other methods of investigation the positivist might devise can pass the scientific test of replicability (whether or not the same results can be found by other scientists using the same methods). Natural laws can be discovered empirically, and once found they must be made subject to verification by other qualified investigators.

Sociologists use quantitative methods in their research to attempt to uncover patterns or relationships between social *variables*. Variables are those factors which might influence what is being studied. They can be a social trait, a quality, or a characteristic that varies or is subject to change. Although there are literally an infinite number of variables, some of those used more commonly in sociology are social class (upper, middle, lower), political affiliation (Democrat, Republican, Independent), religion (Protestant, Catholic, Jew, etc.), and education (years of schooling completed).

Variables are either *dependent* or *independent*. The *dependent variable* is the factor that is being affected, usually a behavior or an attitude. It is so named because it depends on other variables. The variable that influences the dependent variable is the *independent variable*. A period of time exists between the independent and dependent variables, with the former coming before the latter. For example, if sociologists were interested in voting behavior, they would ask people how they voted (Democratic, Republican, Independent, or other) and then would try to understand the criteria or variables which affected voting behavior. Social class can be used as an illustration. By a series of questions sociologists can ascertain to what social class the voter belongs. In this example, how one voted is the dependent variable and social class is the independent variable, because people are members of a social class before they vote. Class affiliation therefore has been used to predict voting behavior.

In order to test the relationship between the dependent and

independent variables another element—the *test factor*—is introduced by the sociologist. The test factor enables the sociologist to discover if the relationship between the dependent variable and the independent variable is a true causal relationship, or what is called a *nonspurious relationship*. In other words, the sociologist asks the question: Does a third variable explain the relationship? Does someone vote Democratic (dependent variable) because he or she is middle class (independent variable) or because he or she is Jewish (one possible test factor). According to accumulated evidence on voting behavior, sociologists claim that the test factor, Jewishness, is a much stronger explanatory variable than is being a member of the middle class. Thus, there is a spurious relationship between voting behavior and social class. However, when Protestantism is used as the test factor, it does not change the relationship between voting behavior and social class. Therefore, a nonspurious relationship exists between voting behavior and social class for Protestants but not for Jews. It then follows that researchers can use an infinite number of test factors to see if indeed a nonspurious relationship has been discovered.

A Model for Quantitative Research

Over the years a consensus has been achieved regarding an ideal research process or model to use when engaging in quantitative research. Depending upon the particular investigator, quantitative research can consist of anywhere from three or four steps to over a dozen. Scimecca and Sherman offer a six-step process that combines the best of the numerous approaches. This model is as follows:

1. *Problem Generation and Specification.* Step one is to conceive of a research project. The research may seek confirmation of a theoretical assumption. Durkheim, in his book on suicide, for example, wanted evidence that suicide rates are related to the spread of social change. Research may also test popular assumptions. For example, are crime rates really rising or just receiving more media attention? Another possible research endeavor might be to determine how effective a new type of self-help group is in reducing alcoholism or drug addition. Thus the source of a study can be a theoretical question that requires evidence, an empirical finding that requires explanation, or a problem faced by an individual or group that might be helped by sociological research.

2. *Reviewing the Literature.* Step two is to find out what is already known about the issue. Talk to experts and lay people, and then go to the library to review previous research. This enables the researcher to understand and evaluate what is known and to determine what further questions need to be asked. It is at the next stage, the research design stage, that "re-search" or "looking again" begins.

3. *Research Design.* Research design involves carefully outlining what is to be studied and deciding how the data will be collected, systematized, and analyzed. The two major quantitative research designs are *survey research* and the *experiment.* Survey research uses a questionnaire or structured interview to collect data from a sample of the population. The experiment involves creating and observing controlled change.

Of the two methods, survey research is by far the most widely used one in quantitative sociology. A survey research design includes a consideration of the population and how it is to be sampled. Usually, limited time and resources make it impractical to observe every possible person or group researchers are interested in knowing about (what sociologists call the "universe of the study"). However, a proper sampling procedure is seen as solving this problem. Whichever design is used, variables are defined and *operationalized* (specified as to how the variable is to be measured), hypotheses are generated, and ways of testing them are formulated. Research design includes deciding whether descriptive or inferential statistics are most appropriate for a particular study. Descriptive statistics tell what has been found. Inferential statistics allow the researchers to decide whether their results are repeatable or due only to chance. They also allow the researcher to generalize from the sample to the population.

4. *Data Collection.* Researchers, having decided a whole array of issues and having made the appropriate choices, then proceed to the next phase of the study: the collection of data by implementing the research design. During this stage they observe, question, interview, mail out questionnaires if they have chosen a survey design, or decide what changes they want to introduce and how to test the effect of these changes if they have chosen the experimental method.

5. *Data Analysis.* Analyzing data involves finding patterns through the development of categories and systematic comparisons. This is where proper statistical analysis, selected in the research design, is applied.

The researchers look at what the data tells them. They turn the numbers into words and describe the relations found or not found and try to determine what this means. An important part of data analysis is drawing conclusions and developing questions for further research. If the data do not disconfirm their theory, it becomes more credible.

6. *Writing the Report and Making It Public.* Making one's research public presents the author's efforts to others for scrutiny. Research, in this way, is transformed from private to public knowledge. It becomes open to criticism and is either accepted or rejected in an open process. The report should include the research design, methods of collection, data analysis, conclusions, suggestions for further research, and at times policy suggestions. Any errors, prejudices, or misinterpretations can be found and pointed out by others. It is this public scrutiny which establishes the validity and reliability of the researcher's efforts. Dissemination makes it possible for the research to be replicated by others with different groups and samples. Researchers can then have more confidence in findings that are carried out in more than one place and time and that produce consistent results.[7]

Quantitative Research Methods: A Critique

If the above were all that the positivist sociologist attempted, there would be little with which one could find fault. However, in their desire to emulate the natural scientist, positivist sociologists have gone well beyond the limitation of establishing relationships between variables. They have substituted sophisticated techniques for fitness between their measurements and reality. Statistical techniques have been refined and sharpened to the point where they have become an end in themselves. Well-known sociological journals rarely publish any findings which do not use the latest methodological technique, regardless of the importance of what is found. (Humanist sociologists who use other methods of research must publish their findings in less well-known journals or start journals of their own.) This emphasis on sophistication in statistical technique has produced a curiosity in social science—*methodology,* or how one conducts research, has become a subfield with its own specialists who spend a great deal of their professional lives studying *how* to study the world.

Positivist sociologists refuse to acknowledge that the objective, external world is only one part of reality and that there is also a subjective reality, which is constructed by thinking, acting individuals. This subjective reality, which the person experiences in everyday life, does not lend itself to statistical analysis. When faced with contradictory or inconclusive results, positivist sociologists fall back upon the belief that this "failure" can be cleared up simply by perfecting the measuring instruments and statistical techniques. Science, for the positivist, comes from perfect tools. Positivist sociologists fail to examine the premises on which their research is based—they simply assume that reality is fixed and is amenable to perfected instruments. This view fails to allow for the freedom of the individual to choose alternatives.

On the other hand, the humanist sociologist assumes that social reality as lived and experienced by individuals does not coincide necessarily with the methods of analysis used by positivist sociologists. Humanist sociologists hold that sometimes the best method of investigating the world of another is to learn the method the other uses to understand the world. Nonsociologists can also be students of human relations; they too can have social theories and conduct investigations to verify them.

Sociologists who use only quantitative methods have produced studies based on their conception of the world, a world seen in purely quantifiable terms. Such a conception may not be that of the people under study, and because narrowly defined problems are more conducive to statistical measurement, the result of such research has been an accumulation of trivia, so much so, that Alfred McClung Lee can ask:

Why is so much sociological fact-gathering and thinking so remote from actual human struggles? Why are so many "scientific" sociological research methods so artificial, so trivializing, so oversimplifying and overcomplicating, so distorting of human affairs?[8]

The answer to Lee's question is that the tail has wagged the dog. Quantitative methods have led positivist sociologists to accumulate an array of minor studies, which are not linked to each other save for the fact that they all use quantitative methods. Because of this, sociology at present is a discipline that finds itself unable to make any empirical generalizations that hold over time or area from the minute studies that fill its journals. Statistical techniques do not mirror social reality no matter how mathematically sophisticated they are. The important questions of freedom, power, emotion, and social dynamics simply are not easily quantifiable. Advocates of quantitative methods have become so caught up in the collection and analysis of their data that they have refused to confront the fundamental contradictions of their methodological stance. They have taken it on faith that the natural science model fits social behavior.

This inadequate view of social reality is reinforced by government agencies—the primary source of funding for sociological research—who, like the rest of the society, are impressed with science and equate science with numbers. All of this adds up to an apparent insecurity of the sociologist, to what Marshall Clinard called "The Sociologist's Quest for Respectability."[9]

In the early twentieth century, in order to establish itself as a discipline worthy of being taught in a university, sociology had to satisfy the criticisms of the "older," more established disciplines. Economics, with its generally accepted claim to scientific status, was to be emulated. Sociology, therefore, had to be "scientific" and "objective" to be considered respectable. Science gave sociology a legitimacy in the intellectual community. Richard LaPiere, who lived through this period, recalls how this quest for legitimacy shaped the foundation of modern sociology.

Well into the 1930s the status of sociology and hence of sociologists was abominable, both within and outside the academic community. The public image of the sociologist was that of a blue-nosed reformer, ever ready to pronounce moral judgments, and against all forms of social conduct. In the universities, sociology was generally thought of as an uneasy mixture of social philosophy and social work. . . . Through the 1920s the department at Chicago was the real center of sociology in the United States [but] . . . the men who were to shape sociology during the 1930s were, for the most part, products of one or two-men departments (e.g., Columbia) of low status within their universities; they were, therefore, to a considerable degree self-trained and without a doctrinaire viewpoint, and they were exceedingly conscious of the low esteem in which sociology was held. Such men, and I was one among them, were determined to prove—at least to themselves—that sociology is a science, that sociologists are not moralists, and that sociology deserves recognition and support comparable to that being given psychology and economics. It was, I think, to this end that toward the end of the 20s, scientific sociology came to be identified with quantitative methods . . . and by the mid-thirties American sociologists were split into two antagonistic camps—the moralists . . . and the scientists.[10]

LaPiere thus sees the development of American sociology more as an effort to establish legitimacy and build respectability than to obtain knowledge. In their haste to become scientific, positivist sociologists have also lost sight of the tradition of the Enlightenment from which social science emerged. The *philosophes* sought to fashion a moral science which would grant the individual "the freest possible choices to change social forms."[11] Positivists rule out the idea of the sociologist as a critic of society by their methodological prohibitions.[12]

Before looking at qualitative methods of research, one other criticism of positivism, that mounted by feminist sociologists, needs stating. Feminists hold that the objectivity postulated by proponents of quantitative methods "is not a truth-seeking principle, but derives from the particularistic experiences of dominant males in capitalist society."[13] Feminists hold that male-dominated quantitative methods claim objectivity because they overlook the feminist ideal of identification with the research subject as a fellow human being, opting instead for a vision of the respondent as an object used as a means for the researcher's ends. Feminists' interests in the subjectivity of the individual thus lend themselves to more qualitative methods of research.

In conclusion, quantitative sociological research is limited for the investigation of social reality. Results from its use do not explain adequately human behavior. Therefore, qualitative methods of social research as an addition and an alternative to quantitative research methods are discussed next.

Qualitative Methods

Qualitative methods of social research can be traced back to the tradition in social science which follows the *verstehen* or sympathetic understanding principle. *Verstehen* as a method of analysis was formulated by Max Weber and is based on the assumption that the subjective study of human behavior—putting oneself imaginatively into the place of the person being studied—is a highly effective means for gaining knowledge about social life.[14] By actively participating in the life of the observed, sociologists can use their introspective abilities to achieve understanding. Human beings think in and act upon symbols, and therefore their behavior can be explained only through the perception and understanding of these symbols.[15]

Sociologists who subscribe to the *verstehen* approach are aware of the differences between *objective* and *subjective* knowledge and act accordingly. By invoking such qualitative methods as participant observation, in-depth interviewing, total participation in the activity being investigated, and so forth as research strategies, the sociologist better obtains firsthand knowledge about the social phenomena under investigation. More than quantitative methods, qualitative methods allow researchers to be close to the data. This enables them to develop conceptual and categorical components of explanation from the data itself, rather than from the rigidly structured and highly quantified techniques that pigeonhole the world into the definition that the researcher has already constructed.[16]

Qualitative methods offset the tendency among positivist sociologists to dehumanize the individual by reducing everything to inventory-like categories. The knowledge necessary to understand human behavior is enmeshed in a complex web of social interaction and is not readily acertainable by purely statistical methods of investigation.

Thus, qualitative methods examine the objective social world via an interpretation derived from the perspective of the individuals under investigation. As such they follow Herbert Blumer's dictum to "respect the nature of the empirical world and organize a methodological stance to reflect that respect."[17]

However, care must be taken to avoid exorbitant claims for qualitative methodology, as mainstream sociologists have done for quantitative methods. It bears repeating that both are needed if we are to begin to grasp the complexity of human behavior. With this caveat in mind, I now turn to the major qualitative methods: (1) unobtrusive measures; (2) comparative life history methods; (3) case studies; and (4) participant observation.

Unobtrusive Measures

Unobtrusive measures of observation of social behavior are methods which directly remove the observer from the behavior being studied. They range from examination of secondary records to the use of hidden recording devices. Webb, Campbell, Schwartz, and Sechrest define unobtrusive measures as follows: "The individual is not aware of being tested, and there is little danger that the act of measurement will itself serve as a force for change in behavior or elicit role-playing that confounds the data."[18]

Three major varieties of unobtrusive measures are *physical trace analysis, archival records,* and *content analysis.*

Physical Trace Analysis. Physical traces and signs left by individuals can impart a great deal of knowledge to the researcher who knows what to look for. For example, wear and tear on library books indicate their use; the notation of replaced tile in a museum points to the popularity of specific exhibits; articles strewn around conference rooms and in wastepaper baskets give clues as to what people have been doing and how often.

An ingenious use of physical trace analysis by a Chicago automobile dealer is reported by Webb et al. The dealer estimated the popularity of different radio stations by having mechanics record the position of the station dial in all cars brought in for service. More than 50,000 dials a year were checked, with less than 20 percent duplication of dials. These data were then used to select radio stations to carry the dealer's advertising.[19]

Physical trace methods are most appropriately used as a strategy for recording the incidence, frequency, and distribution of social acts toward specific objects over time and in various situations. For greater accuracy, different populations in different settings and at different times should be compared. In this way, the researcher can make generalizations beyond the single sample. Ideally, data collected by the physical trace method should be used in combination with information gathered by questionnaires and intensive observation, each acting as a check on the others.

There are a number of obvious advantages in the use of the physical trace method. In particular, its inconspicuousness enables the researcher to circumvent the following problems: the individual's awareness of being measured and the alteration of his or her behavior; selection of individuals for observation; interviewer effects; and bias that might come from the measurement instrument itself. Its weaknesses, which obviate its being used alone, are revealed by the answers to the questions: (1) Have the materials selectively survived, if historical? (2) Have they been selectively deposited, if contemporary? (3) Does the time taken to collect the data justify the results? And (4) Does the fact that the observer lacks sufficient data on the total population preclude generalization beyond the immediate situation?

After weighing the pros and cons of the physical trace method, one can easily agree with Webb et al, who state: "If physical evidence is used in consort with more traditional approaches, the population and content restrictions can be controlled, providing a novel and fruitful avoidance of errors that come from reactivity."[20]

Archival Records. The second form of unobtrusive measures is analysis of public and private archival records. Here, investigators must be particularly careful of bias. Archival records inevitably bear the imprint of the organization that produces them. This is further complicated when changes over time are considered. Changes in personnel will, for instance, produce shifts in language and, possibly, differences in perspectives.

Two other major sources of bias are selective deposit and selective survival. Are the records which have survived truly representative? Did those in charge purposely leave archival records that cast a favorable light upon the organization under study? These are just two of the questions to which a sociologist who uses archival records must give thought. However, those who have used this unobtrusive measure believe that the low cost of acquiring a massive amount of data along with the nonartificiality (compared to the questionnaire and interview) far outweigh its disadvantages. By assuming that there will be bias, archival investigators carefully weigh this bias before they draw conclusions.

Content Analysis. Content analysis, although it involves the counting of key words, topics, or themes that appear in books, magazines, newspapers, novels, television series, advertising campaigns, or such other forms of media as art, is usually considered to be a qualitative technique, because its major focus is on forms and patterns rather than on statistical configuration.

Categories are created, and the frequency with which something fits into a category is tabulated in a systematic manner. Classification schemes may identify certain words, the number of words in a typical sentence, ideas, or treatment of persons in pictures. Erving Goffman, for example, showed how women were seen as being subservient to men in advertisements in major magazines;[21] and Greenberg, Richardson, and Henderson found the same trend in prime-time television programs.[22]

Content analysis has also proven to have some predictive value. Namenwirth, for example, analyzed party platforms in presidential campaigns written in the 1960s and was able to accurately predict the economic recession of the early 1980s.[23] A well-known popular use of content analysis is found in John Naisbitt's best-seller, *Megatrends.*[24] Naisbitt starts with the assumption that newspaper space is limited. Thus, it is necessary to choose what will be covered. Since there are only a limited number of issues that the public is concerned with at any given time, issues and combinations of issues tell us a lot about a society. For example, on any given day one can pick up the paper, look at the front page, and see—

or not see—articles of current interest such as abortion, religion, civil rights, crime, nuclear power plants, drug abuse, and so on. On the basis of ideas that emerge on or drop off the front page and the frequency with which issues appear, Naisbitt claims to be able to tell a good deal about the direction in which American society is going—hence the title *Megatrends.*

Questions that arise over the use of content analysis concern selective sampling on the part of the researcher, the biases of the procedures, and perhaps most important of all, whether the classification schemes are reliable and valid.

The Comparative Life History Methods

In the 1930s and 1940s, under the influence of University of Chicago sociologists Robert Park and Ernest Burgess, the comparative life history method was used quite extensively. With the advent of survey methods and statistical techniques, it was put aside, and the result was a loss for soicological analysis.

The life history approach is an attempt to define the relationship of an individual to a cultural milieu. It can include both biographical and autobiographical documents. Although there is no systematic way of defining life history techniques, the psychologist John Dollard offers the following criteria:

    I. The subject must be viewed as specimen in a cultural series.
    II. The organic motors of action ascribed must be socially relevant.
    III. The peculiar role of the family group in transmitting the culture must be recognized.
    IV. The specific method of elaboration of organic materials into behavior must be shown.
    V. The continuous related character of experience from childhood through adulthood must be stressed.
    VI. The "social situation" must be carefully and continuously specified as a factor.
    VII. The life-history material itself must be organized and conceptualized.[25]

Life history materials may include any record or document that throws light on the subjective behavior of individuals. Three types of life history materials have been distinguished in social science: the *complete,* the *topical,* and the *edited.* All three forms contain the following basic features: "The person's own story of his or her life; the social and cultural situation to which the subject and others see the subject responding; and the sequence of past experiences and situations in the subjects' life."[26]

Complete Life History. The complete life history attempts to encompass the subject's entire life experiences. An excellent example of this type of life history is *Tragic Magic: The Life and Crimes of a Heroin Addict,* written by criminologist Stuart Hills and former heroin addict Ron Santiago.[27] Although Hills refrains from offering his own analysis, editing only for clarity and readability, and lets Santiago tell his own story, the sociological and social policy issues raised by the work are readily apparent. The result is a glimpse into a deviant life-style that few people not in that life-style ever encounter. *Tragic Magic* offers a look at a heroin addict's perception of self, his interaction with others, and the cultural milieu in which he lived out his life as an addict.

Topical Life History. The topical life history differs from the complete life history in that only one phase of the subject's life is presented. An example of this type of research is Carl Klockars' *The Professional Fence.*[28] Here, Klockars presents us with one person's conception of his chosen criminal profession.

Edited Life History. The edited life history may be either topical or complete. Its key feature is the continuing interspersing of comments, explanations, and questions by the investigator. A classic example of a topical life history is Conwell and Sutherland's *The Professional Thief* in which annotations and interstitial passages were added to the life history.[29]

The advantages of the life history method are numerous, especially if the investigator finds an articulate subject. Life histories can provide pictures of people's lives that are extremely difficult to obtain in any other manner. When done well, the private side of life, which is usually hidden from the sociologist, is made available.

The pitfalls of this approach, however, are also numerous. Subjects may try to present their best side, exaggerating what is positive and hiding what is negative. Also, memory is a highly selective thing, and no one is immune to this failing. Distortions are likely to arise at any sequence in the biographical account of the subject. However, because they are immersed in the process of investigation, sociologists will be able to spot and perhaps even correct the inaccuracies. In this way, the life history can be used as a basis against which other data from interviews, questionnaires, and all types of observations can be assessed.

Excellent work using this kind of research has been done by anthropologists in the field of ethnohistory. A fascinating complete life history study of blacks living on Maryland's eastern shore during the late nineteenth and early twentieth century is Shepard Krech's *Praise the Bridge That Carries You Over: The Life of Joseph Sutton.*[30] Krech spent countless hours interviewing a ninety-four-year-old black man, whose ability to recall the events of his past was uncanny. Krech scrupulously checked the information given by Joseph Sutton against census records and historical documents and was able to construct a history of a people and a period that otherwise might have been lost forever.

Case Studies

Perhaps the best definition of the case study method has been offered by Stoecker, who believes that the term should be reserved "for those research projects which attempt to explain the dynamics of a certain historical period of a particular social unit."[31] No one specific technique is used; case studies have relied on histories, personal documents, in-depth interviews, questionnaires, and even statistical records.

Like the comparative life history, the case study method was first used extensively in the 1930s and 1940s by University of Chicago sociologists Robert Park and Ernest Burgess, who urged their doctoral students to go out and talk to people. The result was a number of interesting, well-written, and informative monographs, including *The American Hobo* by Nels Anderson, which studied those who would now be called homeless men,[32] and *The Gold Coast and the Slum* by Harvey W. Zorbaugh, an analysis of contrasting life-styles in two sections of Chicago.[33] More recent examples are Rosabeth Moss Kanter's *Men and Women of the Corporation*, which studied the pressures on executives in a corporation,[34] and Mark Jacob's *Screwing the System and Making It Work*, an analysis of a county juvenile justice organization.[35]

The case study method has been criticized because of the problems stemming from its lack of generalizability (an "N of 1," or only one case) and its problems with researchers' biases. Positivists, in particular, claim that a science needs to discover general laws that cover all cases and that the case study can never meet this criterion. As for biases, the criticism is that the method relies on retrospective (and thereby biased) reports and employs arbitrary interpretations.[36]

Defenders of the case study method counter that individual cases can be compared to test a theory, and that researchers can more confidently generalize if it can be shown that generalizations apply to a diverse array of cases.[37] As for bias, Yin argues that this can be overcome by the researcher's reporting of preliminary findings—perferably in the data collection stage—as a safeguard against bias.[38]

Participant Observation

For the humanist sociologist, of all the qualitative methods available, *participant observation* is the most fruitful for gathering information about social reality. Although there are many definitions of the term "participant observation," a fairly clear one is offered by Bogdan and Taylor. They define participant observation as:

> research characterized by a period of intense social interaction between the researcher and the subjects, in the milieu of the latter. During this period, data are unobtrusively and systematically collected.[39]

Participant observation differs from other types of qualitative research in that observers become part of the group of people they study. In participant observation, researchers interact with those they are studying for an extensive period of time. It is not uncommon for the researcher to live in a community for a number of years in order to know the people being studied and to participate in the group's activities. The data gathered by the participant observer consists of extensive field notes which are written from memory when the researcher is alone, usually immediately after the events have occurred.

Probably the most important reason for humanist sociologists to rely on participant observation as their primary method of analysis is that it enables the researcher to assume that the individuals under study have freedom of choice. As Severyn Bruyn writes:

> Social scientists who study society with a predominantly behavioristic or positivistic perspective of man cannot view concepts such as free will as having any basis in fact. Under such conditions it is easy to see how some social scientists reach the conclusion that free will has no reality. Actually, what is considered "fact" by the traditional scientist is only that which he selectively interprets from his own explanatory model, and the model is only part of a wider factual reality.[40]

Participant observers assume freedom of choice on the part of human beings, and this assumption guides their study. The participant observer is in a position to observe how people go about their daily lives, making decisions both rationally and irrationally, and testing their conceptions of reality against others.

What then are the essential features of the method of participant observation which enable the researcher to make sense of the social milieu? Norman Denzin has listed some of the more important components of the participant observation method, and we will look at these.[41]

Denzin's Framework for Participant Observation

1. *Handling hypotheses.* The observer must continually revise and test hypotheses as the study is being conducted.
2. *Thick descriptions and interpretations versus thin descriptions and interpretations.* A thin description merely describes an act or behavior. A thick description, on the other hand, is interpretative and probes the intentions, motives, meanings, and circumstances of the behavior or action.
3. *Assumptions of the method.* The central assumption of direct observation is that subjects have a thorough knowledge of why they do what they do. This is seen as being more accurate than explanations imposed from outside. The investigator may share in the life and activities of the group under study.
4. *Direct participation in the subject's symbolic world.* Participant observers must try to place themselves in the symbolic world of the subjects to see the meanings subjects attach to their actions. The *verstehen* method requires that participant observers immerse themselves in the subject's reality.
5. *Creating an identity by taking a role in the subject's interactions.* Observers carve out roles for themselves in the ongoing situations under study. They try to blend into the ongoing interactions where and when appropriate.

Participant Observation: A Critique

As a humanist sociologist, I believe that the participant observation method is the best research tool available to the sociologist. However, this should not preclude an examination of the criticisms of the method offered by other sociologists.

The major criticism raised about participant observation concerns its *generalizability.* Stated simply, can participant observers claim that what they have uncovered in one context is applicable to other milieux? The answer is, it depends. Researchers must be aware of whether the population studied has unique characteristics and whether instabilities in the population are such that they limit any conclusion about the sample at hand. In the final analysis, the generalizability of any study depends on whether it can be supported by other similar studies. Participant observation, like quantitative methods, must accept replication as a test. If it cannot be repeated, there is no basis for considering it anything but an idiosyncratic description of a social phenomenon.

Another criticism arises concerning the question of bias inherent in the method. By immersing themselves in the process of observation, do observers lose their objectivity? Humanist sociologists reject this criticism because they believe that objectivity is only an ideal. Quantitative methods of analysis have just as many, if not more, biases. By being aware and making his or her biases explicit, the humanist sociologist is much more honest and accurate in his or her research than is the sociologist who relies only on quantitative methods.

A third criticism relates to the reactive effects of the observer; in any setting, the observer becomes a foreign object. The mere fact that the observer is present alters the situation. A prime example of this type of reactivity is Festinger, Riecken, and Schacter's famous study *When Prophesy Fails.*[42] The presence of observers who had assumed a false identity to gain access to a small religious cult solidified the belief of the group that their prophesy was correct and thereby changed the group under study.

Observers should begin their study with an awareness that produces subsequent checks on just how much reactive effect will be allowed to alter the situation. By accepting the view that bias is present, participant observers are more realistic than sociologists who employ questionnaires and believe that they are being completely objective.

Another criticism concerns "going native," or the changes that may occur in the observer. Perhaps, the best example of this is reported by Kurt and Gladys Lang in their study of a Billy Graham crusade. Two of their participant observers became participants by dropping the role of researcher and walking down the aisle to make their "Decision for Christ."[43] Although this is rare and attests to the persuasive skills of Billy Graham, changes in the observer nevertheless are bound to occur, and the observer should be aware of this process. It is incumbent upon the participant observer to record these changes and deal with them by talking both to informants and colleagues so as to interpret the effect that these changes have on the analysis.

The above criticisms can be minimized through the proper education and training of would-be researchers in sound qualitative methodological techniques. This includes viewing themselves as they would see any other participant in the situation; weighing their own influence when they analyze their data; and giving sufficient detail concerning procedures when they report their findings in order to enable readers to similarly weigh the potential sources of reactivity.[44] Furthermore, when these procedures are taken into consideration, the advantages of the method of participant observation far and away exceed those of the questionnaire and interview. These advantages can be catalogued as follows:

1. The participant observer is not bound by prejudgments about the nature of the problem, by rigid data-gathering devices, or by preconceived hypotheses as is one who uses the questionnaire and survey.

2. The participant observer can avoid using meaningless and irrelevant questions.
3. Participant observers are better able to rely on their own impressions and reactions during the research process than are survey analysts.
4. The participant observer is in an excellent position to go beyond the public selves of the respondents as compared to the interviewer who used a questionnaire and has only a fleeting relationship with the respondent.
5. Participant observers are equipped to link statements and actions to respondents because they are present in the situation where the interaction takes place. The survey analyst who restricts observations to one occasion must rely much more than the participant observer on indirect inferences about behavior and statements.
6. The participant observer is the best-equipped person and is closest to the data as it is being collected. In the survey, the most qualified person is typically the survey director. However, it is not usually the survey director who conducts interviews in the field; this is too frequently left to the homemaker, the part-time student, or the graduate assistant. Thus, in doing a survey, there are a number of levels between the field director and the data. This is not the case in participant observation.
7. The participant observer can make greater use of data from informants and impressionistic reactions than those who use the survey can. The participant observer thus is in a better position to impute motive from observation, to pace observations at a rate that leads to low levels of refusal, and to incorporate what may have at first seemed to be irrelevant data than is the survey analyst.[45]

   In conclusion, participant observation would appear to provide the best way to gain an understanding of the complexity that characterizes human behavior. This method is not without faults, and those who use it should be aware of the distortions produced by it. The chief strength of participant observation is that the researcher is permitted to examine the social world from the perspective of the subject; it enables the researcher to achieve a "subjective reality." Such a reality can take into account the varieties of human behavior, which can help make sense of the world we live in. To this end, qualitative methods in general, and participant observation in particular, can take us further along the path to answering the fundamental question of social science: Why do people behave the way they do?

Ethics in Sociological Research

All social scientific research, no matter what the discipline, raises questions about the ethics of the research—the responsibility of the researcher to the people he or she studies. Although questions concerning values are usually entwined with those of ethics, given that ethics implies the *ought*, this is not a problematic consideration for the humanist sociologist. Value freedom, the view that the sociologist's values do not enter into the research, is dismissed out of hand by the humanist sociologist. The value-free researcher is a myth. For the humanist sociologist, the research act implies that one takes sides for or against particular values. As stated in chapter 1, the definition of humanist sociology used in this work is: *the study of human freedom and of all the social obstacles which must be overcome in order to insure this freedom.* Therefore, the major concern of the humanist sociologist is how the research helps the individual achieve freedom.

What is problematic for the humanist sociologist is the ethics involved in the research process itself. Questions about consent, deception, and harm that all sociologists wrestle with are very much on the minds of humanist sociologists.

The position taken here is that there is no one absolute answer. Concerns about ethics in social research relate to four general questions, and the fourth question is the most important one. The four questions are:

1. Were the subjects deceived by the sociologist?
2. Were the subjects harmed by the study?
3. Did the study violate the privacy of the subjects?
4. Did the study contribute to the freedom of individuals in the society (not necessarily the subjects of the study)? And if so, did the benefits of the study outweigh the first three considerations?[46]

For example, if a sociologist were to study the Ku Klux Klan, a group predicated on racial hatred, one would expect that the sociologist, in order to accumulate reliable date, would have to (1) deceive the subjects, (2) bring reputational harm to the group, and (3) violate the privacy of the group. However, such a study could contribute to the freedom of non-Klan members, and in so doing, the benefits of the study would far outweight the first three considerations.

Unfortunately, determining the answers to these questions in relation to the overwhelming majority of studies is not as easy as it is when studying a deviant hate group such as the Klan. With this in mind, let us look at an actual study that points up certain ethical ambiguities associated with sociological research to see how the four questions might clarify the moral issues.

The study is Laud Humphreys' research on homosexual behavior in public restrooms.[47] Humphreys used participant observation techniques to systematically observe how men who are strangers engage in homosexual encounters in public restrooms, or "tearooms," as they were referred to in the slang of the homosexual world. Although he did not engage in homosexual sex, Humphreys participated through the role of a "watchqueen," a third party. In Humphreys' words:

> Fortunately, the very fear and suspicion of tearoom participants produces a mechanism that makes . . . observation possible: a third man (generally one who obtains voyeuristic pleasure from his duties) serves as a lookout, moving back and forth from door to windows. Such a "watchqueen," as he is labeled in the homosexual argot, coughs when a police car stops nearby or when a stranger approaches. He nods affirmatively when he recognizes a man entering as being a "regular." Having been taught the watchqueen role by a cooperating respondent, I played the part faithfully while observing hundreds of acts of fellatio.[48]

It was the second stage of Humphrey's research that triggered considerable controversy. He recorded the license plates of a number of the tearoom participants and one year later interviewed them in their homes as part of another study in which he was engaged. Humphreys claimed that none of the fifty men he interviewed recognized him because he had changed his appearance. This enabled him to gather demographic data which showed that a large number of these men were married with children. This led to the conclusion that many men who engage in casual sex in "tearooms" did not think of themselves as homosexuals; indeed, they led "respectable" lives in order to hide "the discreditable nature of their secret behavior."[49]

A number of critics felt that Humphreys' behavior was so deceitful that it had no place in sociological research. Columinist Nicholas von Hoffman held that Humphreys "collected information that could be used for blackmail, extortion, and the worst kind of mischief without the knowledge of the people involved."[50] Von Hoffman concluded his argument against Humphreys by stating: "No information is valuable enough to obtain by nipping away at personal liberty, and that is true no matter who's doing the gnawing . . . the conservatives over at the Justice Department or Laud Humphreys and the liberals at the sociology department."[51]

Sociologists Irving Louis Horowitz and Lee Rainwater defended Humphreys by pointing out that he conducted his research in a principled manner and at all times protected the identity of his respondents. To have made a full disclosure to the individuals would have resulted in aborting the project, and the information gained about the particular form of homosexual behavior would have been lost.[52]

As a humanist sociologist, my position is that Humphreys obviously deceived the subjects, did indeed cause harm to them, did violate their privacy and, because it did not contribute to the freedom of others, the benefits of the study did not outweight the first three considerations.

As previously stated, Humphreys' study was a controversial one. Another study, though only tangentially a social scientific study, presents ethical issues that are less complicated. Begun in the 1930s as a research experiment by the United States Public Health Service, this study sought to explain what would happen if syphilis went untreated. Some four-hundred poor Alabama blacks were identified as having syphilis and given placebos. Treatment was withheld from them for more than forty years in order to wait for autopsies to be performed.[53] Here the answers to the four questions are quite clear: this is a study that should never have been attempted.

Summary

This chapter has tried to show the limitations of positivism and its use of quantitative methods in sociological research. Positivism, which presupposes the existence of natural, invariable laws and deterministic interpretation of social behavior, was shown to fall short on two levels: (1) the fundamental premises underlying sociological positivism are not representative of reality, and (2) sociological positivism has produced an exceedingly meager core of knowledge about human behavior.

Qualitative methods were proposed as an alternative to quantitative methods that dominate mainstream sociology. Unobtrusive measures, the comparative life history method, the case study, and participant observation were presented as the principal examples of qualitative methodology. Though not without flaws, participant observation was considered to be the best method available for understanding social interaction.

The question of ethics in sociological research was raised. Four basic questions must be asked concerning the ethics of a study: (1) Were the subjects deceived by the sociologist? (2) Were the subjects harmed by the study? (3) Did the study violate the privacy of the subjects? (4) Did the study contribute to the freedom of individuals in the society (not necessarily the subjects of the study)? And if so, did the benefits of the study outweight the first three considerations?

Notes

1. Auguste Comte, *The Positivist Philosophy*, 3 vols. Harriet Martineau, trans. New York: Calvin Blanchard, 1854/96.

2. Steven Seidman and David G. Wagner, *Postmodernism and Social Theory* (Cambridge, MA: Basil Blackwell, 1992).

3. Jonathan Turner, "The Promise of Positivism," in Seidman and Wagner, *Postmodernism and Social Theory*, pp. 158–62.

4. William I. Thomas, *The Child in America* (New York: Knopf, 1928), p. 584.

5. Anselm Strauss and Juliet Corbin, *Basics of Qualitative Research: Grounded Theory, Procedures and Techniques* (Newbury Park, CA: Sage, 1990).

6. C. Wright Mills, *The Sociological Imagination* (New York: Oxford University Press, 1959).

7. Joseph A. Scimecca and Arnold K. Sherman, *Sociology: Analysis and Application* (Dubuque, IA: Kendall/Hunt, 1992), pp. 28–30.

8. Alfred McClung Lee, *Sociology for People* (Syracuse, NY: Syracuse University Press, 1988), p. 19.

9. Marshall B. Clinard, "The Sociologists' Quest for Respectability," *Sociological Quarterly* 7 (1966): 399–412.

10. Richard LaPiere, personal communication to Irwin Deutscher, cited in Irwin Deutscher, "Words and Deeds: Social Science and Social Policy," *Social Problems* 13(1965):24.

11. Ernest Becker, *The Structure of Evil* (New York: George Braziller, 1968), p. 37.

12. Richard J. Bernstein, *The Reconstruction of Social and Political Theory* (Philadelphia: University of Pennsylvania Press, 1978), p. 53.

13. Patricia M. Lengermann and Jill Nierbrugge-Brantly, "Feminist Sociological Theory: The Near-Future Prospects," in George Ritzer, ed., *Frontiers of Social Theory* (New York: Columbia University Press, 1990), p. 326.

14. See Max Weber, *Max Weber on the Methodology of the Social Sciences*, Edward Shils and Henry Finch, eds. (New York: Free Press, 1949).

15. Herbert Blumer, *Symbolic Interactionism* (Berkeley: University of California Press, 1986).

16. William J. Filstead, "Introduction," *Qualitative Methodology: First-hand Involvement with the Social World*, William J. Filstead, ed. (Chicago, IL: Markham, 1970).

17. Herbert Blumer, *Symbolic Interactionism: Perspective and Method* (Englewood Cliffs, NJ: Prentice-Hall, 1969), p. 60.

18. Eugene J. Webb, Donald T. Campbell, Richard D. Schwartz, and Lee Sechrest, *Unobtrusive Measures: Nonreactive Research in the Social Sciences* (Chicago, IL: Rand McNally, 1966), p. 175.

19. Ibid.

20. Ibid., p. 52.

21. Erving Goffman, *Gender Advertisements* (New York: Harper and Row, 1979).

22. Bradley S. Greenberg, Marcia Richardson, and Laura Henderson, "Trends in Sex-Role Portrayals on Television," in Bradley S. Greenberg, ed., *Life on Television* (Norwood, NJ: Ablex, 1980).

23. J. Z. Namenwirth, "The Wheels of Time and the Interdependence of Value Change," *Journal of Interdisciplinary History* 3(1973):649–83.

24. John Naisbitt, *Megatrends* (New York: Warner Books, 1983); *Megatrends 2000* (New York: Morrow, 1990).

25. John Dollard, *Criteria for the Life History* (Freeport, NY: Books for Libraries Press, 1971), p. 8.

26. Norman Denzin, *The Research Act,* 3d ed. (Englewood Cliffs, NJ: Prentice-Hall, 1989), pp. 187–88.

27. Stuart Hills and Ron Santiago, *Tragic Magic* (Chicago, IL: Nelson-Hall, 1992).

28. Carl B. Klockars, *The Professional Fence* (New York: Free Press, 1975).

29. Chic Conwell and Edwin H. Sutherland, *The Professional Thief* (Chicago, IL: University of Chicago Press, 1937).

30. Shepard Krech, III, *Praise the Bridge That Carries You Over: The Life of Joseph Sutton* (Boston, MA: Schoekman, 1981).

31. Randy Stoecker, "Evaluating and Rethinking the Case Study," *Sociological Review* 34(1991):88–112.

32. Nels Anderson, *The American Hobo: An Autobiography* (Leiden, The Netherlands: Brill, 1975); originally published in 1923.

33. Harvey Zorbaugh, *The Gold Coast and the Slum* (Chicago, IL: University of Chicago Press, 1929).

34. Rosabeth M. Kanter, *Men and Women of the Corporation* (New York: Basic Books, 1977).

35. Mark Jacobs, *Screwing the System and Making It Work* (Chicago, IL: University of Chicago Press, 1992).

36. William M. Runyon, "In defense of the case study method," *American Journal of Orthopsychiatry* 52(1982):440–46.

37. Jennifer Platt, "What Can Case Studies Do?" *Studies in Qualitative Methodology* 1(1988):1–23.

38. Robert K. Yin, *Case Study Research* (Newbury Park, CA: Sage, 1989).

39. Robert Bogdan and Stephan J. Taylor, *Introduction to Qualitative Research Methods* (New York: Wiley, 1975), p. 4.

40. Severyn T. Bruyn, *The Human Perspective in Sociology* (Englewood Cliffs, NJ: Prentice-Hall, 1966), p. 44.

41. The following section is based on Denzin, *The Research Act,* pp. 158–62.

42. Leon Festinger, Henry Riecken, and Stanley Schacter, *When Prophesy Fails* (New York: Harper and Row, 1956).

43. Kurt Lang and Gladys E. Lang, "Decision for Christ: Billy Graham in New York City," in Maurice Stein, Arthur J. Vidich, and David M. White, eds., *Identity and Anxiety* (Glencoe, IL: Free Press, 1960), pp. 415–27.

44. Bogdan and Taylor, *Introduction to Qualitative Research Methods.*

45. Adapted from Denzin, *Research Act,* pp. 169–70.

46. Adapted from Scimecca and Sherman, *Sociology: Analysis and Application,* p. 44.

47. Laud Humphreys, *Tearoom Trade* (Chicago, IL: Aldine, 1970).

48. Ibid., p. 12.

49. Ibid., p. 135.

50. Nicholas von Hoffman, "Sociological Snoopers," *Trans-Action* 7(1970):4,6.

51. Ibid., p. 6.

52. Irving L. Horowitz and Lee Rainwater, "Journalistic Moralizers," *Trans-Action* 7(1970):5–8.

53. James H. Jones, *Bad Blood: The Tuskegee Syphilis Experiment* (New York: Free Press, 1981).

CHAPTER 3

The Development of the Self

The question "Who am I?" is something we all ask ourselves at one time or another in our lives. The search to find oneself, to establish an identity, has been a major concern of human beings since thought processes first developed. If a humanist sociology is to offer answers to this question, it should provide some insight into the process of how individuals come to view themselves, or how the self develops.

The humanist sociological view, like the traditional sociological view, begins with the assumption that human beings are born into a particular society and acquire its culture. As individuals grow from help-less infants to adults, they become functioning members of society—they become social persons. The individual is, in a word, *socialized*. Social-ization, as defined here, implies that the individual acquires a commit-ment to the norms, values, beliefs, and behavioral patterns of his or her society. This commitment occurs because he or she internalizes the norms and, in the process, comes to accept the rightness of applying a particular norm or norms to a specific interaction or situation.[1]

To internalize means to incorporate the norms into one's internal patterns of behavior. At an early age, children will refrain from engaging in an action they have been told not to do if their parent or some other authority figure is present and they believe that they will be admonished or punished in some way if caught. Later, children learn to refrain from the action even though no one will know they have done it. The popular expression for this is that "one's conscience is one's guide"; the sociolo-gist's expression is that "one has internalized the norms."

Sociologists and anthropologists hold that the individual's self de-velops within the context of a group that has established rules and regulations. Individuals also learn to act differently in different situations and at different times of their lives. Norms change; we make choices. Socialization is a lifelong process. We change as the situations in our lives change. Socialization implies adaptation as we make and remake our lives.

The process of socialization and the development of self varies within and between different societies. We cannot develop as social beings nor participate in a society unless we know the rules of the particular society, unless we are socialized. Without socialization, the distinctively social characteristic of the self would not emerge. It is only through socialization that the individual acquires a personality, an organization of beliefs, habits, and behavior.[2]

To be considered fully human, an individual must be a member of a society or at least have some contact with a member of a society. Human beings need emotional attachment with at least one other person. Without such attachment there is no socialization, and the personality does not fully develop. Evidence in support of this is given by accounts of feral children and others who have been raised in isolation. Because there have been few such cases, generalization is difficult. Nevertheless, a clear picture does emerge, one that supports the need for interaction between self and society.

Feral Children

The word *feral* means "untamed," and there are a number of cases of children who were discovered living in the wild, having had contact only with wolves and other animals. Two of the most famous cases of feral children are Victor, the Wild Boy of Aveyron, reported in 1799 by Jean Itard,[3] and Kamala and Amala, the wolf children of India, reported in 1920 by the Reverend J. A. L. Singh.[4]

In 1799, hunters found a seventeen-year-old boy living in the woods like a wild creature, surviving on nuts and berries. Victor, as the boy came to be called, walked, raced, and climbed on all fours. Diagnosed as an incurable idiot by leading French psychologists, he was entrusted to the care of Jean Itard, a medical doctor and a teacher in a special school for the deaf and mute. Although Itard taught Victor some human mannerisms, he learned only three words before he died at the age of forty.

Two girls, Kamala and Amala, were eight and one-and-a-half years old, respectively, when the Reverend Singh took them into his orphanage in India. Like Victor of Aveyron, they had previously lived among wolves. Amala died one year later, but Kamala lived to the age of seventeen;

although she always ran on all fours, she did learn to walk erect. Before she died, Kamala had mastered about fifty words.

Children Raised in Isolation

Two important cases of children who were raised in isolation, Anna and Isabelle, were reported by Kingsley Davis.[5] More recently, Susan Curtiss and Russ Rymer reported on the case of Genie.[6]

Anna was discovered at the age of six. Anna's mother was unwed, and Anna's grandfather forced her to hide Anna in the attic so as not to bring shame to their family. Anna received just the bare minimum of physical care and had little if any social interaction. When Anna was found, she could not walk, talk, or feed herself; she could not take care of herself at all. At first, she was thought to be deaf and blind. Anna had missed nearly six years of socialization. Her condition showed how little purely biological resources, when acting alone, contribute to making someone a complete person.[7]

Clinicians and social workers attempted to socialize Anna but with little success. She was able to learn some words but not sentences. Her motor development resembled that of a two-year-old. She learned to walk but ran clumsily. She died before she reached eleven years of age, and we can only speculate as to what might have happened to her in later life.

The second case reported by Davis is that of Isabelle. She, too, was discovered when she was about six years of age, and like Anna was the child of an unwed mother. Isabelle's grandfather had insisted that she be kept in a dark room. However, unlike the case of Anna, Isabelle's mother, a deafmute, stayed with her. Isabelle thus differed from Anna in that she interacted with and was shown emotion from another person, her mother. Since her mother was a deafmute, Isabelle had no chance to hear spoken language, but she did learn to communicate with her mother through the use of gestures.

Isabelle made great strides and by the time she was eight-and-a-half reached an apparently normal level of development, so much so that she could attend school with children of her own age. The early interaction with another human being, her mother in this case, seems to have made the difference between her development and that of Anna.

Genie was discovered at thirteen years of age. From the age of one-and-a-half she had been locked in a small room, tied to an infant's potty chair. Her deranged father beat her whenever she made any sounds, and she never developed any language capability. When she was discovered by the authorities and admitted to a hospital, she was diagnosed as having the mental age of a one-year-old. She could not walk properly or chew food. Her only spoken words were "stopit" and "nomore."[8] Specialists at UCLA designed a learning program for her, and she became the focus of acrimonious battles in the courts between educational specialists, mental health professionals, friends, and family. However, she made little if any progress in development.[9] Obviously, we can not infer too much from the cases presented here. However, they do lend some credence to the sociological perspective on human development. Our self—our personality—develops only in interaction with others; humanness is a social product that results from socialization. Of the cases cited, only Isabelle, who had contact with another human being (even though it was minimal), developed any semblance of what would be considered a normal self.

Socialization transforms a biological organism into a social person. The self develops only in a social context. There is a give-and-take relationship between the self, roles, institutions, institutional orders, and society. Each is defined by the others and reciprocally defines them. Due to this interaction, humans are both creatures and creators of society. This integrative process, this stressing of the interrelationship between the self and the society, constitutes a sociological understanding of the way personality or self develops. However, it should be pointed out that the sociological view on the development of the self is not the major view taught in the colleges and universities of the United States nor the major view held by the general public. The most popular views are psychological, with behaviorism still the leading view among academic psychologists and Freudianism the most popular with the public. Therefore, these two major psychological views on the development of the self are presented here. Because both behaviorism and Freudianism have a tendency toward determinism, they are critiqued in light of a humanist perspective. The sociological perspective of George Herbert Mead and the thoughts of the anthropologist Ernest Becker are then offered as the basis of a humanist theory of the development of the self.

Skinner And Behavioral Theories

During the 1920's, positivism, with its emphasis upon observable fact, began to prevail in the social sciences. Theories of self which dealt with internal states were labeled "armchair theorizing" and condemned as unscientific. Behaviorism, which emphasized that only measurable behavior was worthy of study, became inseparable from positivism and was spread by such influential psychologists as J. B. Watson, Lynn Thorndike, Clark Hull, and later by perhaps the most widely known behaviorist of all, B. F. Skinner.

Behaviorism proved to be extremely attractive to psychologists who wanted to establish psychology as a science rather than leave it as a branch of philosophy. Theories which could not be supported by demonstrable facts were relegated to the fringes of the scientific community. Any concept which could not be proven by the empirical method was deemed unworthy of scientific consideration. Because the notion of self referred to an internal activity and did not lend itself to measurement, it was of little use to the early behaviorists.

Behaviorists offered statistical results based on unbiased, controlled laboratory experiments which were believed to be more scientific than results based on single individuals who merely examined their own consciousness. And because all forms of life could be examined using the scientific method, behaviorists felt that rats and pigeons, which were easier to study, could be substituted for human beings.

In behaviorism, the psyche, the mind, and consciousness are left to the philosophers, discarded because they cannot be observed. As J. B. Watson, the father of behaviorism, stated over three-quarters of a century ago: "The time has come when psychology must discard all reference to consciousness."[10] This approach has not changed very much. And over fifty years later, in a work written to bring behaviorism to the attention of the general public, B. F. Skinner wrote:

> We can follow the path taken by physics and biology by turning directly to the relation between behavior and the environment and neglecting supposed mediating states of mind. Physics did not advance by looking more closely at the jubilance of a falling body, or biology by looking at the nature of vital spirits, and we do not need to try to discover what personalities, states of mind, feelings, traits of character, plans, purposes, intentions or other prerequisites of autonomous man really are in order to get on with a scientific analysis of behavior.[11]

Behaviorism rejects the notion of a reflective mind and assumes that human beings are passive and inert beings whose actions are determined by external forces. Humans can no more influence their own destinies than can a robot and are thus susceptible to prediction.

Behaviorists study only behavior constructed and generated in a controlled experimental environment. They design research strategies that seek results based on controllable, observable, and quantifiable variables.[12] If human beings are the subjects in the experiment, they are placed in situations of subservience and powerlessness. The possibility of breaking through this mechanistic model is not allowed, and the behaviorist then is able to claim that freedom does not exist. This line of reasoning becomes apparent in the thought of a contemporary behaviorist, B. F. Skinner.

B. F. Skinner (1904–1989)

Skinner's influence in psychology can be traced to his first published volume, *The Behavior of Organisms*.[13] However, it is with two of
his popular works, *Walden Two*[14] and *Beyond Freedom and Dignity*,[15]
that his influence extended beyond the psychological community to the
general public.

In all of his works, Skinner insisted on the positivistic method of
studying human behavior. The concept of mind and the purpose of
behavior have no role in Skinner's conception of science. Skinner is a
thoroughgoing determinist. Frazier, the central character in Skinner's
novel *Walden Two*, states: "I deny that freedom exists at all. I must deny
it—or my program would be absurd. You can't have a science about a
subject matter which hops about capriciously."[16]

Laws of Reinforcement and Extinction

Skinner's theories of behavior might be called the *laws of reinforcement and extinction*. Briefly, Skinner believed that behavior is influenced
by changes in the environment. If a behavior results in favorable consequences, the *law of positive reinforcement* is at work, and individuals will
engage in that behavior again in order to receive the reward. However,
should a behavior result in unfavorable consequences, then the *law of
negative reinforcement* comes into play, and individuals will not repeat
that behavior. When a decrease in the frequency of a behavior occurs,
the *law of extinction* is in force, which means that a previously reinforced
response has lost its power because of a lack of reinforcement.

Such theories of behavior can lead to several problems. First, there
is the problem of what constitutes a reinforcer. According to Skinner,
"the only defining characteristic of a reinforcing stimulus is that it reinforces."[17] Skinner argues that this is not circular reasoning, that there
is nothing wrong with classifying events in terms of their effects. But
his argument is unsound. To accept his explanation is similar to accepting
the primitive notions that rain gods exist because it rains and that war
gods are real because there are wars.

Second, Skinner and the behaviorists fall into the same trap as do
positivist sociologists—they assume that the environment or objective
reality they believe exists (in this case, a deterministic one) is the *same* as
that created by those individuals being studied. Behavior is not examined
from the point of view of the subjects under study. (How do they feel
about what they are doing? How do they interpret their actions?) The
only *meaning* attributed to the behavior under study and the only social
context considered is that which the behaviorist deems real.[18]

Another criticism of Skinnerian behaviorism is that outside the laboratory the studies are open to question. For example, Breland and Breland, two early supporters of Skinnerian behaviorism, eventually came to believe that:

> Our first report ... concerning our experiences in controlling animal behavior was wholly affirmative and optimistic, saying in essence that the principles derived from the laboratory could be applied to the extensive control of behavior under non-laboratory conditions throughout a considerable segment on the phylogenetic scale. When we began this work, it was our aim to see if the science would work beyond the laboratory, to determine if animal psychology could stand on its own two feet as an engineering discipline ... However, in this cavalier extrapolation (from the Skinner box), we have run afoul of a persistent pattern of discomforting failures. These failures, although disconcertingly frequent and seemingly diverse, fall into a very interesting pattern. They all represent breakdowns of conditioned operant behavior.[19]

One more criticism of Skinnerian behaviorism is that behaviorists come very close to embracing political authoritarianism. If the human organism is as completely controlled as the behaviorists hold, then there is nothing that can be done about the future. To make this state of affairs a bit more palpable, however, the behaviorist shifts the analysis to the society or culture. If individuals are automations without value, then it is the society which should survive. Skinner states it this way: "Survival is the only value according to which a culture is eventually to be judged, and any practice that furthers survival has survival value by definition."[20] Therefore, individuals exist solely to help the culture survive. To put this into political terms, individuals exist for the state. The possibility that some societies should not survive (Nazi Germany comes immediately to mind) is never raised by Skinner.

Behaviorists have negated the moral and political primacy of the human person, disavowing responsibility and freedom for the individual and placing more importance on society, culture, or whatever term is given to external determining forces. Indeed, as Floyd Matson eloquently states:

> If it is true that man "does not act upon the world, the world acts upon him"; if there is nothing inside a man's skin different (except in degree of accessibility) from what is outside it; if there is no potential for responsibility, resourcefulness, and reflection—in short, if there is no freedom for man—then it is not too much to say that man himself has been explained away and in his place there grins the image of the cheerful robot.[21]

Therefore, a humanist sociology must reject behaviorism for two important reasons: (1) it is an erroneous conception of human behavior, and (2) it is a potentially authoritarian political doctrine.

Freud and His Theory of the Self

The story of Sigmund Freud (1856–1939) and his theories of personality and behavior have been described and discussed in numerous works and therefore will not be covered here except in brief and then only as they pertain to the concept of the self.

Freud thought that the personality consisted of three systems—the *id*, the *ego*, and the *superego*—which function in the manner of the body's physical system (keep in mind that Freud's early training was as a physician).

The Id

The id is the unconscious part of the personality and is present from birth. This psychic reality can be said to represent the inner world of subjective experience. Some theorists consider it to be the representation of the baser instincts of human beings, giving rise to behavior that is characterized as blind, irrational, or even brutish. The id also is that part of the personality which seeks immediate gratification as a response to frustrations of behavior. This is called the pleasure principle.

The Ego

The ego is the cognitive system of the personality and can be characterized as being responsible for perception, decision, and thought in human beings. The ego might be called the executive of personality because it controls action; that is, it selects the features of the environment to which it will respond. The ego is the organizing system of the personality; it exists to carry out the aims of the id, and its powers are derived from the id. Freud stated:

> The ego must on the whole carry out the id's intentions; it fulfills its task by finding out the circumstances in which those intentions can best be achieved. The ego's relation to the id might be compared with that of a rider to his horse. The horse supplies the locomotive energy, while the rider has the privilege of deciding on the goal of guiding the powerful animal's movement. But only too often there arises between the ego and the id the not precisely ideal situation of the rider being obliged to guide the horse along the path by which it itself wants to go.[22]

The ego thus has no existence apart from the id and is never completely independent of it. The principal function of the ego is to mediate between the instincts of the person and the surrounding environment.

The Superego

The third system of the personality is the superego. To Freud, the superego was the internalization of the repository of values and ideals, or belief systems, of the society. It is the moral aspect or conscience of the self. The superego becomes the moral aspect of the self and must decide right or wrong. According to Hall and Lindzey, the superego has three main functions: (1) to inhibit the impulses of the id, particularly those of a sexual or aggressive nature, since the expression of these impulses is highly condemned by society, (2) to persuade the ego to substitute moral goals for realistic ones, and (3) to strive for perfection. Therefore, the superego decides what is right and wrong in order to act in accordance with the moral standards authorized by society.[23]

In sum, the id, ego, and superego are how Freud saw the human personality. These three segments work together as a unit under the guidance of the ego to allow the individual to function as a whole. In a very general way, the id can be seen as the biological, the ego as the psychological, and the superego as the social system of the personality.

Although Freud's theories of personality formation are popular, they have a somewhat limited application to a humanistic theory of self based on these two reasons: (1) Freudianism offers a deterministic view of human behavior, and (2) Freudianism is politically conservative and thereby often supports a repressive status quo.

Frued, Determinism, and Political Conservatism

Freud was trained as a physician and came to intellectual maturity during the influence of the deterministic and positivistic philosophical traditions of the late nineteenth century. He accepted the scientific notion that the human organism is a complex system of energy which is used in such processes as circulation, respiration, and muscular exercise. This physiological energy also is transformed into psychic energy—exemplified by the id, the ego, and the superego—and thus determines human behavior.[24]

Such a theory of behavior can give rise to a political conservatism which overlooks the inequalities in modern society. Freud thought the root of human problems lay not in the unequal distribution of power and possessions but in individuals' instinctual nature, which is at odds with society; the problem is within the individual and not within the society. Therefore, individuals must conform, or adapt themselves, to society. Such a doctrine discourages change in society. A humanist sociology cannot accept an idea such as this which overlooks the possibility of freedom of behavior and therefore the possibility of change.

On the other hand, another twentieth-century view, that of Meadian social-psychology, holds much for a humanist theory of self. As shall be seen, it offers a picture of a dynamic, active individual, one who "acts upon the world" rather than being a slave to external forces.

Mead and Social-Psychological Theories

Median social-psychology is based on the theories of George Herbert Mead. Mead called his view of the development of the self *social behaviorism* to distinguish it from the behaviorism of J. B. Watson. To avoid confusion, the term social-psychology will be used to refer to Mead's theories.

As described in chapter 1, Mead believed that all group life was based on cooperation between individuals. Human beings create a social world, unlike animals, whose world is only biological. For example, after spending their mature years in the open sea, salmon complete their life cycle by swimming upstream to their place of birth, where they spawn and die. Unlike animal behavior, human behavior is not determined by instinct but is shaped by social interaction.

Social Interaction

To accomplish social interaction, individuals have to be able to predict what others will do in order to guide their own behavior accordingly. The ability to predict the behavior of others is based on the capacity to imagine what it is like to be the other person. The use of imagination as part of the cognitive process is what differentiates between human and animal behavior. Unlike animals, humans can choose how and to what they will react. Human behavior, then, is not simply a matter of responding to external forces or to the activities of others; "rather it involves responding to the intentions of others, i.e., to the future, intended behavior of others—not merely to their present actions."[25]

Human response is based on the interpretation of gestures. Clues are provided by gestures which enable the participants to imagine the thoughts and future actions of each other, an act called *role taking*. Cooperative activity can begin when individuals participating in concerted activities are able to attach the same meaning to the same gesture, which becomes what Mead called a *significant symbol*. The process of agreement upon and development of significant symbols involves *meaning* and depends on the use of language. The crucial importance of language in this process should be stressed.

> It is through language (significant symbols) that the child acquires the meanings and definitions of those around him. By learning the symbols of his group, he comes to internalize the definitions of events or things, including the definition of his own conduct.[26]

Social Order

Mead thus builds his theories of behavior on the assumption that there is a social order consisting of the responses humans develop to each other and then postulates that the self develops within this context. Of paramount importance is the ability to get outside one's self and to be able to take the role of another. Of course, role taking is simply a metaphor for the projection of one's self into another's situation and imagining the feelings of the other. And while these imagined feelings may be quite real and intense, it should be kept in mind that these feelings are one's own and not the other person's. Because of this limitation, the accuracy of one's role taking always has about it a certain amount of uncertainty.[27]

The meaning of the social nature of the self can be more fully seen by reviewing what Mead considered to be the stages of self-development in the child: (1) the preparatory stage, (2) the play stage, and (3) the game stage.

The Preparatory Stage

The earliest stage of the self is the preparatory stage. In this stage, the child imitates the adult, and there is no real symbolic understanding taking place. As yet the self is not defined.

The Play Stage

The second or play stage develops when the child begins to acquire language. As children learn language they are able to define those words that have shared meaning. It is also at this stage that the child comes to assume the perspective of those whom psychiatrist Henry Stack Sullivan called *significant others* or those who are important to the child and from whom the child seeks acceptance.[28] These significant others become what the sociologist refers to as role models, individuals from whom children learn to regulate their behavior. To the small child, significant others are responsible for the emergence of self; children come to view themselves as objects because of significant others. To some extent, individuals necessarily see themselves as they believe significant others see them.

At this stage, the child has not developed the capacity to see him- or herself from the perspective of more than one person at any one time. This is the reason why Mead calls this the play stage. As Joel Charon writes:

> Play is an individual affair, subject to the rules of single individuals. Mead's play stage is a time when the child takes the roles of significant other—

father, Superman, mother, teacher—and acts in the world as if he or she were these individuals. In taking the role of these others the child acts toward objects in the world as they act, and that includes acting toward self as they do. This stage is the real beginning of the self as a social object.[29]

The Game Stage

The third or game stage represents organization and with it the necessity of assuming the perspectives of several others simultaneously. Group life and cooperation depend upon knowing one's position in relationship to large numbers of individuals and to a complex set of rules. According to Mead, the incorporation of the attitudes of significant others and the rules and regulations one is exposed to produces a *generalized other*.

> The play antedates the game. For in a game there is a regulated procedure, and rules. The child must not only take the role of the other, as he does in the play, but he must assume the various roles of all participants in the game, and govern his action accordingly. If he plays first base, it is as the one to whom the ball will be thrown from the right field or from the catcher. Their organized reactions to him he has embodied in his own playing of the different positions, and this organized reaction becomes what I have called the "generalized other" that accompanies and controls his conduct. And it is this generalized other in his experience which provides him with a self.[30]

What Mead called the generalized other eventually expands to include the values and expectations of the entire community. Knowledge of the perceived expectations—what the community deems important—of the generalized other establishes an important link between the fully developed self and society. This self, with the incorporated generalized other, is composed of two distinct parts: the "I" and the "Me."

The "I" and the "Me"

Mead believed that the I represents the individual's impulsive tendencies and the Me, on the other hand, represents the attitudes the individual has incorporated from others.

In this conception, each act begins with the I and ends with the Me. The I is the spontaneous initiation of action before the self comes under the control of the expectations of others or the regulatory Me.[31] It is from this give-and-take that novelty and nonconformity arise. Mead stated:

The "I" is something that is never calculable. The "Me" does call for a certain sort of "I" insofar as we meet the obligations that are given in conduct itself, but the "I" is always something different from what the situation itself calls for. So there is always the distinction, if you like, between the "I" and the "Me." The "I" both calls out the "Me" and responds to it. Taken together they constitute a personality as it appears in social experience. The self is essentially a social process going on with these two distinguishable places. If it did not have these two places, there could not be conscious responsibility and there would be nothing novel in experience.[32]

The formation of the I and the Me has definite implications for a humanist conception of self in that it offers a picture of an individual who possesses an active mental life or mind. What Mead called *minded behavior* is the self in action. To solve a particular problem, people think about the different possibilities or alternatives for future action. The individual envisions the future and then selects from among the possibilities. This selection process implies the *constructing* of acts rather than mere responding in a predetermined way.[33]

Mead offered a conception of human beings who actively determine their own actions rather than merely passively respond to external stimuli. Individuals make both themselves and their society.

Like the self, the mind is social. Through association with others, people internalize the definitions of language, assume the role of others, and come to think. Because individuals possess a mind, they can maintain and adjust themselves in society, which enables the society to persist. Thus, the persistence of society depends upon consensus, and consensus depends upon individuals possessing minds.[34]

Bernard Meltzer has summarized the process of the development of self and mind as follows:

The human individual is born into a society characterized by symbolic interaction. The use of significant symbols by those around him enables him to pass from the conversation of gestures which involves direct, unmeaningful response to the overt acts of others—to the occasional taking of the roles of others. This role-taking enables him to share the perspective of others. Concurrent with role-taking, the self develops, i.e., the capacity to act toward oneself. Action toward oneself comes to take the form of viewing oneself from the standpoint, or perspective, of the generalized other (the composite representative of others, of society, within the individual), which implies defining one's behavior in terms of the expectations of others. In the process of such viewing of oneself, the individual must carry on symbolic interaction with himself, involving an internal conversation between his impulsive aspect (the "I") and the incorporated perspective of others (the "Me"). The mind, or mental activity, is present in behavior whenever such symbolic interaction goes on—whether the individual is merely "thinking" (in the everyday source of the word) or is also interacting with another individual. (In both cases the individual must indicate things to himself.)[35]

While Mead's theories of the development of the self are important from a humanist perspective, Mead failed to develop an adequate notion of emotion and motives or a dynamic theory of the affective life of the person. Nor did he explain fully individual diversity or idiosyncratic behavior. He showed how the self develops socially through the generalized other, but he placed too great an emphasis on how it mirrors the outside world. Consequently, though he raised the possibility of novelty and nonconformity in human behavior, he did not adequately deal with the internal discords of individuals. It is for these reasons that we now turn to the writings of Ernest Becker, who sought to combine the thought of Freud and Mead and is only now receiving his due, to see if his approach to the development of the self takes us further along the road to a humanist theory of the self.

Ernest Becker and the Development of the Self

Ernest Becker (1924–1974) was an anthropologist who developed a social theory of human beings as the animals in nature who impose symbolic categories of thought on raw experience. Society, in Becker's view, is a game, a play form. In playing at society, human beings create a symbolic world which gives meaning to their lives. This ability to create a symbolic world enables individuals to transcend the limits of their instincts.[36]

Becker's contribution to a humanist theory of self was his combination of Meadian and Freudian concepts about the social development of self and how human beings overcome their biological limitations. For example, Becker believed that the early training of children is best understood not through the Freudian belief in the Oedipus and Electra complex (the view that children are erotically attached to the parent of the opposite sex) but through what Becker labeled the "principle of self-esteem maintenance." In a long passage Becker explained how this principle works:

> The whole early training period of the child can be understood in one simple way: it is the period in which he learns to maintain his self-esteem by avoiding anxiety, and the anxiety comes from his human environment in the form of disapprobation or the threat of separation from the parents. Thus, he maintains his self-esteem precisely by forming himself into the type of person who need not fear disapprobation or the loss of his succoring objects. . . . This means that he becomes human by learning to derive his self-esteem from symbolic performances pleasing to the adults rather than from continued psychological dependence, which becomes displeasing to

them. The various styles of human character (or life styles) which result from the early training can then be considered as variations in modes of self-esteem maintanance. Thus, in the most brief and direct manner, we have a law of human development and its explanatory principle.[37]

In short, individuals simply like to feel good about themselves.

Becker was in sympathy with both Freud and Mead, but he went beyond them. One problem with Mead's theory of self is its assumption that each person has similar self-conceptions because each is exposed to a similar generalized other. But Mead had difficulty explaining the rupture of social relations. Freud, on the other hand, focused on the process of social disruption at the level of the family. Freud's theory of how the ego develops showed how individuals use different techniques to ward off anxiety while trying to maintain their self-esteem. It is this behavior that produces social friction.[38] The problem is that by relying so heavily on instinct in the form of the id, Freud implied a more or less fixed basic human nature. Freud made a biological problem out of a sociological one.[39]

Becker rejected Freud's emphasis upon instinct, holding that the child has no innate aggressive and sexual drives but only a need for closeness, continuing affection, and protection.[40] Building on this, Becker challenged Freud's views on the unconscious:

> The unconscious now refers simply to the fact that in his early training each child is formed into a particular world view. This kind of exclusive training leads him to distort or obliterate certain perceptions—perceptions not pleasing to his parents, or taboo in the particular society in which he is brought up. Today we speak of the unconscious in quite matter-of-fact terms: it has lost its mystery. Instead of being a fatal subterranean core which we inherited from Paleozoic times, it refers to the particular skewing of our world view and our capacity to act, which occurred during our early training or mistraining. Man is not saddled with a phylogenic fate, but rather with his own early choices, which are designed by his parents—by their tyranny, impatience, or simply their own limited world view. In other words, with the overthrow of instinct theory, Freud's biological problem has again been reconverted to a social and historical one.[41]

Because Becker criticized both Freud and Mead, one should not be misled into thinking that their theories have no value. Freud showed how human beings are helpless and given to anxiety. Mead showed how we learn expectations, values, and rules from our society. Self-feelings and the attempt to avoid anxiety thus depend on the social milieu in which one grows up.

Instead of the neutral term *self-feeling*, Becker used the term *self-esteem*. Thus, high or low self-esteem is related to the individual's ability to navigate safely in society. Human beings tend to feel good about themselves when they come to have a firm command of the conditions that seek to limit their feelings of self-worth.[42] Becker further theorized about mental illness and thought it might be caused by constrictions of self-power. Individuals who are unable to maintain self-satisfaction in the face of new and problematic situations become mentally ill.[43] Schizophrenia, for example, could arise from the inability of the person to take the present firmly in hand and create a future.

Becker on Socialization

Individuals learn from their parents to anchor their self-conceptions in a narrow range of approved choices. In numerous instances the child is not free to choose from a broad range and quality of new experiences. Individuals become rigid from a rigid pattern of socialization. Socialization—the formation of human beings out of helpless, dependent animal matter through the internalization of the expectations, values, and goals of their society—explains the original formation of the social self.

In American society, socialization takes the form of restriction rather than growth, of limitation rather than potential. Adaptation to the demands of parents often can cripple the ego's theoretically limitless potential. Defense mechanisms arise from these restrictions and become techniques of self-deception.[44]

Self-image thus is vital to a humanist theory of self. It is at the very core of human adaptation, growing out of the individual's efforts to deal with anxiety. The development of the ego is summed up in the qualitative feeling of self-value, which is basic to human actions. It is the key to understanding the experiences of socialization.

The early behavior of children centers on their learning how best to maintain self-esteem. A child quickly learns that parental approval (which at first is synonymous with self-esteem) is based on the child's conformity to specific behavior patterns which are understood through symbols or language. As Becker eloquently phrased it, the child's "vital sentiment of self-value no longer derives from his mother's milk, but from his mother's mouth."[45] Self-esteem does not take root in the biological sphere but in the internalized expectations of society as filtered through the child's parents. Children's feelings of self-worth therefore depend on language, for this is how children learn what is expected of them. It is now that the child has been socialized, has been humanized. Human beings are the only creatures in nature who depend on a symbolic constitution of their worth.[46]

Becker and a Humanist Theory of Self

Becker's theory of self can be summed up as follows: An individual's self-formation is built on a denial of the anxiety that arises from the individual's smallness and helplessness as a child for a comparatively long period and on the child's fear of loss of support from his or her parents. It is natural that the child should not want to remain small and helpless. Therefore, the symbolic self becomes the means of changing from weakness to strength, and the personality develops as a vehicle for this change.[47]

By combining Mead's theories of a social self with Freud's theories of anxiety-ridden beings, Becker provides a truly dynamic theory of self. Now a humanist theory of self—one that seeks to integrate self and society—can be fashioned. In order to do this, we must go beyond Becker as he went beyond Mead and Freud.

Becker was a social phenomenologist, one who holds that the social world is solely a product of the ideas that individuals hold of it. He believed that merely understanding the nature of social reality, of what he referred to as its "fictional status," would be sufficient to change this reality. He assumed that if society is the product of individuals' minds, then the course to follow in order to change it would be to change people's views of it. In other words, if humans create society, they can change society.

Becker's position was similar to that of the radicals of the late 1960s and early 1970s who held that the "revolution is in your head." Unfortunately, the process of changing reality is much more complex. George Ritzer states:

> People construct social reality and that reality comes, in turn, to play a role in the social creation of people. . . . The social structures that our ancestors have created historically came to have an existence of their own, an existence that is, at least for the historical movement, beyond the control of people who created them in the first place. We are constrained, if not controlled, by these . . . structures.[48]

Society is an objective reality, a reified (thought of as having life) entity. Social definitions are produced within a political context and are upheld by interests with an immense stake in their preservation. It is this objective reality and how it is organized and structured that we will be concerned with in the remainder of this book. How the self relates to this objective reality is one of the questions raised in chapter 4, where roles, institutions, and social organization are examined. Subsequent chapters deal with the form social organization takes in our society and the political, economic, and social interests which uphold this organization. The role that a humanist sociology can play in understanding the relationship between self and society and how it can help individuals to overcome the forces that limit their freedom and autonomy will be stressed.

Summary

In this chapter, the focus has been on theories about the development of the self, including: (1) Skinner and behavioral theories; (2) Freud and his theory of the self; (3) Mead and social-psychology theories; and (4) Becker and his phenomonological view of the self.

Behaviorism, with its purely deterministic view of self and mind, was dismissed as having nothing to offer to a humanist theory of self. Although both Mead and Freud were shown to have limitations, they both developed important insights into the formation of self. What is needed is a theory of self that overcomes their limitations and builds upon their insights, a post-Meadian, post-Freudian view of self-development. Ernest Becker's views on self, particularly his principle of self-esteem maintenance, were offered as providing a dynamic, integrated theory of self, which build upon the foundation left by Mead and Freud. Becker was seen as not dealing fully enough with social structure, something which will be the focus of the remainder of this text.

Notes

1. William J. Goode, "Norm Commitment and Conformity to Role-Status Obligations," in Bruce J. Biddle and Edwin J. Thomas, eds., *Role Theory: Concepts and Research* (New York: Wiley, 1963), p. 313.

2. Daniel E. Hebding and Leonard Glick, *Introduction to Sociology* (Reading, MA: Addison-Wesley, 1981), p. 81.

3. Jean Itard, *The Wild Boy of Aveyron* (Englewood Cliffs, NJ: Prentice-Hall, 1962).

4. Joseph Singh and Robert Zingg, *Wolf-Children and Feral Man* (Hamden, CT: Archon Books, 1966).

5. Kingsley Davis, "Extreme Social Isolation of a Child," *American Journal of Sociology*, 45(1940):554–64; Kingsley Davis, "Final Note on a Case of Extreme Isolation," *American Journal of Sociology* 52(1947):432–37.

6. Susan Curtiss, *Genie: A Psychological Study of a Modern Day "Wild Child"* (New York: Academic Press, 1977); Russ Rymer, *Genie: An Abused Child's Flight from Neglect* (New York: HarperCollins, 1993).

7. Davis, "Final Note on a Case of Extreme Isolation."

8. Rymer, *Genie*.

9. Ibid.

10. J. B. Watson, "Psychology as the Behaviorist Views It,'" *Psychological Review* 20(1913):163.

11. B. F. Skinner, *Beyond Freedom and Dignity* (New York: Bantam Books, 1971);12–13.

12. Braginsky and Braginsky, *Mainstream Psychology: A Critique* (New York: Holt, Rinehart and Winston, 1974).

13. B. F. Skinner, *The Behavior of Organism: An Experimental Analysis* (New York: Appleton-Century, 1938).

14. B. F. Skinner, *Walden Two* (New York: Macmillan, 1948).

15. Skinner, *Beyond Freedom and Dignity*.

16. Skinner, *Walden Two*, p. 257.

17. B. F. Skinner, *Science and Human Behavior* (New York: Macmillan, 1953), p. 72.

18. Braginsky and Braginsky, *Mainstream Psychology*, pp. 67–68.

19. Keller Breland and Marian Breland, "The Misbehavior of Organisms," *American Psychologist* 16(1961):684.

20. Skinner, *Beyond Freedom and Dignity*, p. 130.

21. Floyd W. Matson, *The Idea of Man* (New York: Delacorte Press, 1976), p. 128.

22. Sigmund Freud, *New Introductory Lectures on Psychoanalysis*, James Strachey, ed. and trans. (New York: Norton, 1965), p. 77.

23. Calvin S. Hall and Gardner Lindzey, *Theories of Personality* (New York: Wiley, 1966), p. 35.

24. Ibid., p. 36.

25. Bernard N. Meltzer, "Mead's Social Psychology," in Jerome G. Manis and Bernard N. Meltzer, *Symbolic Interactionism: A Reader in Social Psychology*, 2d ed. (Boston, MA: Allyn and Bacon, 1972), p 6

26. George Herbert Mead, *Mind, Self and Society*, Vol. 1, Charles W. Morris, ed. (Chicago, IL: University of Chicago Press, 1974), p. 45.

27. George J. McCall and J. L. Simmons, *Identities and Interactions*, rev. ed. (New York: Free Press, 1978), p. 131.

28. Although Mead is usually given credit for first using this term, he never explicitly did so. Henry Stack Sullivan was the first to use "significant others." See his *Conceptions of Modern Psychiatry* (New York: Norton, 1953).

29. Joel M. Charon, *Symbolic Interactionism* (Englewood Cliffs, NJ: Prentice-Hall, 1979), p. 67.

30. George Herbert Mead, "The Genesis of Self and Social Control," *International Journal of Ethics* 35(1925):269.

31. Meltzer, "Mead's Social Psychology."

32. George Herbert Mead, *George Herbert Mead on Social Psychology*, Anselm Strauss, ed. (Chicago, IL: University of Chicago Press, 1956), p. 233.

33. Meltzer, "Mead's Social Psychology."

34. Ibid.

35. Ibid., pp. 17–18.

36. Ernest Becker, *Beyond Alienation* (New York: George Braziller, 1967).

37. Ernest Becker, *The Structure of Evil* (New York: George Braziller, 1968), p. 328.

38. Ibid., pp. 149–50.

39. Ibid., p. 153.

40. Ibid., p. 154.

41. Ibid.

42. Ibid., p. 157.

43. Ernest Becker, *The Birth and Death of Meaning,* 2d ed. (New York: Free Press, 1974), p. 151.

44. Ibid., pp. 56–57.

45. Ibid., p. 67.

46. Ibid.

47. Ibid., pp. 142–43.

48. George Ritzer, "Toward an Integrated Sociological Paradigm," in William E. Snizek, Ellsworth R. Fuhrman, and Michael K. Miller, eds., *Contemporary Issues in Theory and Research: A Metasociological Perspective* (Westport, CT: Greenwood Press, 1979), p. 34.

CHAPTER 4

Social Organization: Society, Culture, and the Role of Roles

Human beings develop a self because they are born into a society and interact with other people. Our lives are lived out in relation to what sociologists call *social organization*, the routines or patterns that develop among people over time. Sociologists, humanist or otherwise, study these patterns, how they are created, and how they influence human behavior.

As people interact with each other, behavior becomes more organized and less spontaneous. This occurs because they need to have an idea of how another person will act in order to act accordingly. We are all in the same situation; thus, patterns are created which remove the element of surprise. Without these patterns, life would be intolerable. Imagine what it would be like if each time you entered a classroom there was no way of knowing what your instructor might do. What if the instructor turned a fire hose on the class, or jumped on the desk and screamed obscenities, or sat silently in a corner, thumb in mouth? It would probably not be long until that instructor was removed from the classroom. Nevertheless, we can see the need for expected, patterned behavior.

Harold Garfinkel founded a branch of sociology called *ethnomethodology*, which seeks to understand the process of how individuals organize and pattern their reality. He devised a number of experiments which show just how important social organization is to the individual.[1] Garfinkel had his students do such things as act as boarders in their own homes; question everything that was said to them in the course of a day; and enter a store and argue with the salesperson over the price of an item. A typical result was the following student report:

On Friday night my husband and I were watching televison. My husband
remarked that he was tired. I asked, "How are you tired? Physically, men-
tally, or just bored?"
(H) I don't know, I guess physically mainly.
(W) You mean that your muscles ache or your bones?
(H) I guess so. Don't be so technical.
(After more watching)
(H) All these old movies have the same kind of old iron bedstead in them.
(W) What do you mean? Do you mean all old movies, or some of them,
or just the ones you have seen?
(H) What's the matter with you? You know what I mean.
(W) I wish you would be more specific.
(H) You know what I mean![2]

As can be seen, people have to understand each other's meanings
and intentions in order to function. In Garfinkel's example, the husband
stated "You know what I mean" twice, becoming more and more frus-
trated until he finally told his wife to "drop dead." His frustration increased as he began to think that she did not know what he meant.
Individuals must develop routines of behavior and establish conditions
whereby they are less and less surprised by each other's actions, where
each feels they know what the other "means." When this consensus
occurs over time, social organization is present.

The two most important types of social organization are *culture*
and *society*. Although these two terms are often used interchangeably,
it is possible to distinguish between them. A word of caution should be
inserted here. Although culture and society can be analyzed separately,
given their intrinsic relation to the individual the distinction is often
blurred in everyday life. With this in mind, then, we can define culture
as consisting of the ideas, language, values, goals, rules, and possessions
shared by members of a society. A culture can also be shared by smaller
units in a society, and when this occurs it is called a *subculture*. Italian-
Americans share a subculture, as do the Daughters of the American
Revolution and major league baseball players. A society, on the other
hand, is one of the largest entities that sociologists study and is equated
in this work with the nation-state.[3] Society is seen as consisting of the
people who share a common culture and the structures or patterns
they create.

Culture

Sociologists do not use the term *culture* as it is used in the popular sense—
appreciation of the arts, literature, music, or the theater. To them, everyone who lives in a society possesses culture. Sociologist Robert Bierstedt

has provided a useful definition of culture: "Culture is the complex whole that consists of all the ways we think and do and everything we have as members of a society."[4] *Thinking, doing,* and *having* give us the three major components of culture—*ideas, norms,* and *things.* Ideas and norms are part of the nonmaterial aspects of culture, and things are part of the material culture, what Bierstedt calls *materiel.*[5] We can now examine each component in more detail.

Ideas

Ideas are a major part of any culture. They include language, beliefs, values, and goals, to name the most important types.

Language. Symbolic language enables us to communicate with each other. As pointed out earlier, George Herbert Mead was the first scholar to systematically emphasize that the development of self could not occur without language. Indeed, humanist sociologists would argue that it is language that distinguishes human beings from animals. Through the use of language humans accumulate knowledge, learn from each other, then use this learning as a basis for future growth.

Language is the mechanism through which children learn to convey and retain meaning and thus develop the ability to think in abstract terms or to move beyond the immediate situation. It is also through the acquisition of language that children become capable of reflection—the ability to think about past experience, to reflect upon it, and to integrate it into a comprehensive view of reality. Present experiences and future activities are combined in the imagination, and through this process, children become conscious of themselves as persons who can reflect upon the relationship of the world to themselves. This is the literal meaning of *reflection,* or *turning back* from the outside world to oneself. Language thus holds the key to the development of reflective, critical human beings.

Beliefs. Beliefs are views about what is true in the world. Every culture contains numerous shared statements about what is true or false, and these beliefs form the foundation for the particular culture. The United States is characterized by the belief in a single God, and the Island of Bali in Indonesia is characterized by a belief in multiple gods. However, not everyone in a society accepts all the cultural beliefs. Not every American believes in the existence of God, and not every Balinese believes in many gods.

Those individuals who deviate from the accepted beliefs are exposed to a great deal of pressure to accept the dominant beliefs. Galileo, who believed that the earth was not the center of the universe, was considered deviant and was tried as a heretic. So important are the shared beliefs of a culture that those who deviate from them are often labeled as insane or mad. Such labeling strips these people of legitimacy and puts to rest the possibility that they may be right. The sentencing of political dissidents in the former Soviet Union to terms in mental institutions is a prime example of how far this has been carried.

Values. A value is what a culture considers important. It is something to which the culture commits itself. In every situation, what a person does depends on what that person values. Competition is an example of a value. The view that competition is not a part of human nature but is a learned value is difficult for some students to accept.

For example, the Hopi Indians were found to be reluctant to stand out from the group. Hopi do not compare their achievements or the importance of their work. A highly skilled stonecutter is perfectly content to earn the same salary as an unskilled laborer. Teachers in Hopi schools have reported that Hopi children break out in tears when singled out for public praise. Hopi children simply do not compete in school, in class work, or even in playing games. One school reported that the children learned to play basketball easily and loved to play the game but could not be taught to keep score. They would play by the hour without knowing or caring who was winning.[6]

People are not born with values; values are learned. Every group emphasizes some values at the expense of others, which gives rise to a conflict of values in a society. For example, although Americans value equal opportunity, a conflict of values occurs when this equality is extended to homosexuals teaching in the public schools or serving in the armed forces. Clashes over values have even resulted in violence, as witnessed by the conflict between those who favor a pro-life stance and those who favor a pro-choice stance toward abortion. Nevertheless, even with this conflicting value system, Americans, on a very general level, do share a common set of values that differentiate us as a people from the cultures of other countries, such as France, England, or Spain.

Goals. Goals differ from values in that they are more practical. Values also have a moral connotation to them which goals do not. Goals are specific. A soccer team wants to win the state cup; a real-estate salesperson wants to join the million-dollar sales club; a student wants to be a college graduate. All of these are goals.

Goals are related to values in that values provide the framework for goal achievement. Hence, the value of competition and winning is at the heart of the soccer team's quest; material success guides the real-estate salesperson; and a belief in the benefit of education pushes the student to meet the requirements for graduation.

Norms

The second component of culture is standards of behavior, or norms. Norms are agreements about what is generally expected of members of a particular culture and are based on the shared expectations of individuals. Examples of norms in American society range from shaking hands upon meeting someone for the first time to wearing clothes in public to not killing fellow human beings.

Norms define how people *ought* to behave in diverse situations. A commitment to the norms of our society is acquired as individuals become socialized. Indeed, socialization, as commonly defined, implies that individuals internalize the norms and in the process come to accept the rightness of applying a particular norm or norms to a specific interaction or situation. People thus learn to act differently in different situations and in different stages of their lives. Eventually, people conform to norms readily, so much so that they become aware of norms only when someone violates them. For instance, Americans would be surprised and even embarrassed if their dinner guests belched after finishing a meal, because such an action would violate a norm. But a belch after a meal is a commonplace occurrence in Japan. When the Thonga of Africa first saw Europeans kissing, they laughed, remarking, "Look at them—they eat each other's saliva."[7]

By offering guidelines for people's behavior, norms ensure that social life runs smoothly. Norms are so important that great emphasis is placed on conformity. Though most people conform to most norms most of the time, they still manage to violate certain norms. In 1906, William Graham Sumner made a still useful classification of norms into folkways, mores, and laws, in relation to how much leeway is permitted for violation of norms.[8]

Folkways. Sumner called the least important norms *folkways.* The term literally means "way of the folk." Folkways are the customs that people follow. Conformity is expected, but if a folkway is violated, people are only considered peculiar not immoral or criminal. If someone constantly picks his or her nose in public, people might eventually come to feel uncomfortable and avoid that person, but there is little that can be done if the nose picker chooses to continue this behavior.

Mores. Mores (pronounced "mor-ays") are much stronger norms than folkways, and a violation is considered immoral. Mores may or may not be written down, but everyone is expected to know them and to conform accordingly. In the United States, cannibalism is so onerous and so obviously a violation of the mores that most states do not consider it necessary to have written laws against it. Incest, on the other hand, is regulated by state laws. Incest is a taboo in all societies, but the ancient Hawaiian royal families expected the ruling line to emerge from brother and sister marriages and did not define this as incest.

The manner in which mores are enforced also varies from society to society. The Trobrianders of the South Pacific considered adultery to be an extremely serious offense and expected offenders to commit suicide. In American society, adultery is not condoned, but adulterers are not prosecuted even though adultery is against the law in many states.

Laws. Unlike folkways and mores, the enforcement of *laws* is formal. Laws are written rules governing conduct which are enforced by governmental authorities and include sanctions and punishments for violating the laws.

For laws to be an effective form of social control, they should also be supported by mores. So-called Blue Laws, which prohibit businesses from opening on Sunday, are carryovers from a time when religion was a much more powerful part of our culture. They are virtually unenforceable today.

In sum, norms are essential elements of a culture, providing people with expectations of behavior and helping to keep social life running smoothly. Norms, along with ideas, comprise the nonmaterial part of culture. We will now look at the third major component of culture, the physical part, or materiel.

Materiel

Materiel is the most tangible and most obvious aspect of culture and includes all of the material items that members of a society have and use. It is impossible to inventory all of the material things of even a small, primitive society, let alone a large industrial twentieth-century society. What is important about the material part of culture is that people learn to use the things that are available in a society and adjust their lives to them. Americans have structured their existence around the automobile, and a threat to the supply of oil by Iraq was seen as such a danger to our national interests that it led to the 1991 Persian Gulf war. One of the most severe punishments parents can threaten their children with is not allowing them to watch television. People born in the last half-century find it hard to imagine what life was like without television.

Other illustrations of how dependent upon material items a society can become are easy to find. A society that has clocks and watches in its material culture exhibits a different tempo of life from one that does not. Societal changes produced by technological inventions are almost beyond reckoning, and their ramifications can seldom be predicted in detail. To cite some obvious examples: What effect has television had on the political process? How has the birth control pill impacted on our sexual mores? How have personal computers changed work habits? In short, the material part of culture affects everything in our lives.

Socialization and Cultural Values

Culture is learned; it is not something we are born with. We learn about our culture as we become socialized. Socialization is a continuous process, beginning at birth and stopping only with death or mental incapacitation. Socialization occurs as humans interact with other human beings and with agents of socialization who teach them about their culture. Specific agents of socialization depend on the person's age and circumstances. For the baby, the socialization agents are the parents, who communicate accepted modes of behavior. Later, peers become agents. Still later, colleagues on the job, spouses, and even such indirect influences as the mass media also come to play an important role in the socialization process.

Socialization occurs at two levels, *primary socialization* and *secondary socialization*. According to Peter Berger and Thomas Luckmann, primary socialization is the initial socialization of the child, the process through which one becomes a member of society. Secondary socialization concerns the processes that primary socialized individuals encounter as they function in the social world.[9]

Primary socialization begins with the family. Because individuals have little choice in the matter, they adopt their family's view of the world. When individuals develop a self and construct a world view for themselves, primary socialization ends. Secondary socialization begins where primary socialization ends and comes into play whenever the individual decides to try something new, whether it be a job, a hobby, or a spouse.

Another kind of socialization is *anticipatory socialization*. This occurs when individuals accept and adopt the values, beliefs, or viewpoints of a group to which they wish to belong but do not as yet belong. Young interns who take on what they perceive as the manner of a doctor and high school graduates who purchase a new wardrobe the summer before entering college are examples of anticipatory socialization at work.

Still another kind of socialization is *re-socialization*. Here individuals are expected to learn a whole new set of values, beliefs, or viewpoints that are different from their own. Marine bootcamp, where civilians are taught to be individuals who follow orders and who would not hesitate to kill in a combat situation, is a prime example of re-socialization.

Socialization thus encompasses the acceptance of the various norms, values, and expectations individuals are exposed to as they become functioning members of society, or social persons.

It is easy to think of socialization as solely a shaping and molding process, and traditional sociologists have fallen into this trap. People are indeed shaped by their culture and molded in such a way as to enable them to participate in the everyday life of the society. But it is important to keep in mind the reflective capabilities of human beings. Even very young children are not passive victims of socialization. Socialization is a reciprocal process; not only the socialized but also the socializers are affected by it. This can be observed in everyday life. Although parents usually succeed to a greater or lesser degree in shaping their children in accordance with the values and expectations that they and the society hold, the parents themselves are changed by the experience. Parents look at the world differently after the birth of their children. Sacrifices are made; different values become important as they seek to provide security for their children. Children's capacity for reciprocity—their capability to act back upon the world and other people who inhabit the world—increases in direct relation to their ability to use language. Children begin to *talk back* to adults, and in the process both are changed.

Society

A society consists of people who share a common culture and is created by their structures or patterns of action. These actions are based on each person's position in the social structure, and people act toward each other according to these positions. Students do not typically address professors by their first name, patients do not usually argue with physicians, and so on. Part of the socialization process is learning one's place or position, what is expected of one, and what is expected of those with whom one interacts. All these expectations comprise the society or social structure.

The main patterns of expectations are *statuses, roles, groups, institutions,* and *institutional orders.* The interrelationships of these patterns form the social structure of a society.

Statuses

Sociologists give *status* two different definitions. The first is synonymous with prestige. A justice of the Supreme Court has high status in American society, but an unemployed high school dropout has low status. (The prestige definition of status is explored more fully in chapter 5, where social inequality is discussed.) The second definition is "the position in society occupied by individuals." Norms and expectations are not "free floating" in a society but are tied to statuses. Society, therefore, can be seen as a network of statuses.

From the moment we leave the confines of our place of residence, our interactions are usually ones of status rather than personal. In other words, our interactions are with the position of the other person and not the person. In the course of a day, students can engage in social interactions with cafeteria attendants, cashiers, bus drivers, salespersons, registrars, and instructors without knowing anything about these people except their status. Social organization makes it possible for interactions to proceed smoothly, and awareness arises only if a situation occurs in which statuses are mistaken or unknown. Many people have experienced being in a store and mistaking another customer for a salesperson. After some slight embarrassment, a quick apology is usually offered for the "mistake." Television comedies often have plots based on "mistaken identity"; the sociologically correct way to say this is that cases of "mistaken status" are a staple.

Statuses are divided into two kinds: *ascribed* and *achieved*. Ascribed statuses are those with which we are born— for example, race and gender. With the exception of those who undergo a sex change operation, these statuses remain until we die. Mary Jones is white, female, and twenty-one years of age. Michael Brown is black, male, and twenty-two years of age. Of these statuses, only age will change through their lives.

Achieved statuses are those that are chosen or earned. One graduates from college, gets married, becomes a lawyer, becomes a parent. All statuses have an important component attached to them—roles. Roles are the personal aspect of status—what the individual brings to it.

Roles

Roles are the dynamic personal parts of statuses—what individuals do in the statuses they occupy. Roles also carry expectations with them. Professors are expected to prepare for a lecture; students are expected to raise their hands before speaking in class. Individuals, though, are given leeway in how they act out their roles. A professor can use alternative methods of teaching, and some classes can be more informal than others. Because of the expectations attached to roles and the orientation be ad hoc. By having a history, institutions become *objective*. For an institution to survive, it must have an existence beyond its creation. Thus, institutions are "experienced as possessing a reality of their own, a reality that confronts the individual as an external and coercive fact."[14] This is called *reification*, or treating concepts and things as if they had a life of their own. Because institutions are treated as having a life of their own, people come to believe that they truly do. Santa Claus provides an example of how reification works. Parents and children treat the institution of Santa Claus as real, and because of this children's behavior becomes controlled and predictable (at least a few weeks before Christmas). Thus, the institution takes on a life of its own regardless of the truth or falsity of its original premise.

As institutions become reified they also take on *legitimacy*; that is, they are seen as being right and proper, and people feel an obligation to conform to the established set of procedures. Legitimations are learned by each succeeding generation and are an integral part of the socialization process. Deviance from institutions becomes harder and harder as they are reified and legitimated. The more conduct is institutionalized, reified, and legitimated, the more predictable and controlled it becomes. Socialization is geared to producing individuals who adhere to the institutions of their particular society.

Here it is important to bear in mind that all societies are pluralistic. Though members of a society may share a common core of values, expectations, beliefs, and so on, there are alternative systems available. Socialization is never complete; there are always ways and means to deviate, and individuals constantly find ways to do so.

One more concept must be introduced in order to understand institutions, that of *social relationships*.

Social Relationships

The notion of social relations is important for an understanding of institutions. Because of reification and legitimation, institutions are seen as being unchangeable. What is often overlooked is that they remain stable only as long as individuals accept them. The German sociologist Max Weber pointed this out in his distinction between individual and social action. Individual action is social action when it takes account of other people's behavior. Making breakfast for oneself is an individual action but making it to share with someone else is social action.[15]

Social relations are based on the probability that one person can understand another's behavior. If someone asks for cream for his or her coffee, he or she does not expect to receive orange juice. For Weber, society is the probability that there will be forms of understandable social action or institutionalized conduct. Society is conceived of in terms of the sum total of probable patterns of behavior.

Social behavior becomes patterned to meet the expectations of individuals. These patterned actions are then reified into a legitimate social order. Social organization exists when individual behavior is determined by rules which are accepted as binding. Institutions are social relationships that are based on the probability that actions will take place according to expectations. The stability of a society is dependent upon reification and perceptions of legitimacy by its members.

Institutional Orders

Institutional orders are those institutions in a society that have similar consequences or ends or that serve similar functions. Sociologists disagree on which institutions should be considered as major orders, but the following five are generally thought to have great importance in any society: political, economic, family, religious, and educational. In modern industrialized societies (the United States being a prime example), a sixth institutional order has arisen—that of the mass media. These institutional orders can be defined as follows:

1. The political institutional order consists of those institutions within which individuals acquire, wield, or influence the distribution of power and authority.
2. The economic institutional order is made up of those institutions in which individuals organize labor, resources, and technical implements in order to produce goods and services.
3. The kinship (family) institutional order is made up of institutions which regulate and facilitate sexual intercourse, procreation, and the early rearing of children.
4. The religious institutional order is composed of institutions in which individuals organize the collective worship of God or deities, usually on regular occasions and at fixed places.
5. The educational institutional order is made up of those institutions where knowledge of the culture is passed on through generations. In modern industrialized societies, the experience of attending some kind of school defines the educational institutional order.
6. The mass media institutional order consists of the institutions in which individuals disseminate news and entertainment to the general public.[16]

The form a particular society takes depends on the interrelationship of the institutional orders. Certain orders dominate and define the characteristics of the society. The United States is dominated by the political and economic institutional orders, and because of this, reality becomes structured in a specific way. The six institutional orders will be examined fully in chapters 6 and 7.

The relationship of the individual to his or her society can be summarized as follows: Human beings are socialized into roles, which carry with them expectations of behavior. These roles are related to institutions, which are reified and legitimated. Major institutions which serve similar functions are referred to as institutional orders. Different types of societies are produced by different arrangements of the institutional orders.

Roles are the key element linking the self to society; they represent the individual's primary contact with social organization. Because of this connecting function, roles and role analysis take on a great importance for humanist sociology. If individual freedom is a reality in a society, it must be found in the link between the individual and the society or in one's roles.

The Role of Roles

The use of the term *role* has a long and varied history. Originally derived from the Latin word *rotula* ("a little wheel, or a round log"), it was used to designate the round wooden roll on which sheets of parchment were fastened to form a scroll. It eventually came to be applied to the script from which sixteenth- and seventeenth-century actors read. The word underwent further changes and came to mean the particular part the actor was playing.[17]

In the 1930s, the concept of role was incorporated into the language of the social sciences by George Herbert Mead and anthropologist Ralph Linton. Mead's work was introduced in chapter 3. Linton's conception of role is discussed here.

Ralph Linton on Roles

In 1936, Ralph Linton offered what has become the best-known social scientific conception of role: he drew the distinction between status as a position and role as the dynamic aspect of status.[18] Linton's conceptions of status and role were extremely influential, and most modern social scientists still accept his insistence upon the close relationship between them. The primary reason for this acceptance is that Linton offered a palpable link between the individual and society. Statuses and their attendant roles are the elements that make up a society.

However, in expanding upon his distinction between status and

role, Linton embraced a deterministic viewpoint, thus setting the stage
for the advancement of this position among mainstream sociologists. In
trying to discover the link between individuals and their culture (Linton
did not distinguish between culture and society), he assumed that everyone agreed on the expected patterns of behavior. Any observed differences between status and role were explained away as some sort of deviance on the part of the individual or by under- or mal-socialization on
the part of those agencies whose charge it is to socialize the human being.
The view that the individual could choose between role expectations was
overlooked by Linton. This is still the dominant perspective in what has
come to be loosely called *role-theory*.

Traditional Role-Theory

Role-theory is not a theory in the sense that structural-
functionalism, conflict, or symbolic interactionism are. Rather, it is a
framework within which certain assumptions and viewpoints are held
in common by social scientists. Role-theory might be more precisely
described as "a theoretical framework concerning roles."

Traditional role-theory starts with the assumption that the expectations of society are transmitted to the individual through the internalization of norms. Norms are the link between roles and behavior. According
to Orville Brim, a well-known student of socialization, the overwhelming
majority of socialization studies begin by asking the "fundamental question of how it is possible for a society to endure and continue to develop.
The inquiry at all times is concerned with how society changes the natural
man, not how man changes his society."[19]

The acquisition of roles is thought to be the most important aspect
of socialization, and because of this emphasis, the dynamic process is
overlooked. Individuals are envisioned by mainstream sociologists as
learning behavior appropriate to their status. An active self is not considered to be necessary for adequate role performance. Individuals capable
of actively shaping their roles are rarely mentioned in the literature on
role-theory. What role-theory emphasizes is the view that human beings
are determined by their social environment. The concept of role is used
to explain social action by linking it to external determination through
the expectations of the other. The role comes to stand for the societal
order (accumulation of statuses), which is then internalized by the individual. Status and role are seen as being so similar that it becomes difficult
to explain why people do not always conform to the expectations attached
to their status-role. This viewpoint is widely accepted, and few sociologists have challenged it. Some possible exceptions to this viewpoint are:
the dramaturgical approach of Erving Goffman; the processual approach

of Ralph Turner; Richard Hilbert's view of roles as an organizing process; and the recent post-modernist views of role seen in the work of Kenneth Gergen and Rose Coser. We will now look at these views to see if, indeed, they do provide exceptions to traditional, deterministic role theory and, therefore, might have relevance for a humanist sociology.

Erving Goffman and the Dramaturgical Image in Role-Theory

The image of the theater is obvious in the literature on role-theory and is usually associated with the name of Erving Goffman[20] Goffman draws an analogy between the players on the stage and the individual (the actor) in society. Just as stage performers have clearly defined parts to play, so too is the actor in society given a clearly defined status or position. Performers follow a written script, and actors in society follow norms; performers follow the dictates of a director, actors obey the wishes of those in power.[21] While on-stage, performers must take into consideration the actions of their fellow performers, and individuals in society must mutually adjust to the responses of others.

For Goffman, there are frontstage and backstage areas that correspond to actual places and to social norms.[22] Individuals prepare for their parts in the backstage area. They get ready to present an idealized version of themselves for public scrutiny: "a performer tends to conceal or underplay those activities, facts and motives which are incompatible with an idealized version of himself and his products."[23] The backstage area is extremely important, and people respect it because it allows for a brief haven where doctors can take off their white coats; corporate executives can put their feet up on their desks; and college students can listen to rock music. Tact protects the backstage region, and as Goffman notes: "Individuals voluntarily stay away from regions into which they have not been invited."[24]

For Goffman, the self is a determined product of the social situation. The primary task of the actor is to manage impressions. Examples of impression management range from the political candidate who is more concerned with the image he or she conveys on television than with the issues in the campaign to the high school student who does not want to appear to be too intelligent lest his or her peers disapprove.

Although Goffman's insights are highly original and his scheme has been applied to all sorts of individuals and situations,[25] in the end he does not offer much for a humanistic theory of roles. The self is reduced to a cynical role player whose primary motivation is one of manipulation. Furthermore, the notion of an active self in role acquisition (the taking of a role) is downplayed in Goffman's writings. The degree of interaction between the person and his or her role is also underestimated.

Although Goffman seemingly offers the actors choices, these choices are played out in a manipulative manner. The individual is so intent on "saving face" that he or she becomes a "con artist." There is no "authentic self" behind the actor's mask. Goffman stresses the ways in which social situations impose rights, duties, and obligations on individuals and how individuals conform to this process. The imposition of external constraints is overemphasized, with implicit changes seen as taking place only in the process of conforming to a role. The major emphasis is upon persons modifying themselves to fit their roles, not on how one might modify roles to fit the person. The self and the role, for all intents and purposes, are the same thing. Goffman in the end opts for a deterministic view of roles.

Ralph H. Turner and the Processual View of Roles

Ralph Turner has long asserted that he is critical of the overly structured, traditional conception of roles in sociology. Turner's position is that the usual conception of role presents a conformity model, one with an overemphasis on normative expectations. Actors in a given status or position perceive what is expected of them, act on these perceptions, and gain social approval for their conforming behavior.[26] For Turner, such a view accounts for a limited picture of behavior. Social interaction involves active construction of conduct among individuals. This is based on reciprocal lines of conduct as individuals seek to come to terms with each other. Taking this into account, Turner extends Mead's concept of "role-taking" to include "role-making," a process in which human beings actively *make* their roles and in so doing communicate this to others. Individuals are faced with loosely defined social situations and in order to interact must make or define a role for themselves. As they do this, they assume that others are also making roles and try to discover just how they do so. Both, then, are called on to give off clues as to what they are actually doing. Role-taking is transformed into role-making and, to Turner, becomes the underlying basis for all human interaction; it is this role-making which enables individuals ultimately to interact and cooperate with each other.

Consistency and not conformity is the crucial element for interaction and understanding of interaction. Individuals must make sense of the other's behavior, and therefore a trade-off in consistency arises. There is an implicit assumption that individuals and those they interact with are consistent in their actions. Consistency is a norm; without it, interactions would be extremely difficult. Interaction is thus "always a tentative process, a process of continuously testing the conception one has of the role of the other."[27] Human beings constantly interpret clues emitted by others as they strive to establish a consistent pattern in the actions of others. Roles are something to be verified in action, and once they are validated, interactions can continue.

The process relates to the self, according to Turner, in that human beings develop self-attitudes and feelings from their interactions with others. Like most role theorists, Turner emphasizes that individuals seek to present themselves in ways that will reinforce their self-image, or in Ernest Becker's view, individuals try to present an image of self about which they can feel good. Because human beings continually seek to determine others' roles, it becomes necessary to inform others, through cues and gestures, of the degree to which their self is anchored in the roles they are playing. Individuals thus consistently inform each other about their self-identity and the extent to which their self is anchored in the role.

Although Turner seemingly focuses upon the dynamic aspects of role behavior by emphasizing role-making rather than role-taking, in his later works he, too, falls into a deterministic trap when he begins to use the term *person*. Person, for Turner, is a repertoire of roles. Attitudes and behavior that develop as an expression of one role carrying over to other situations are, for Turner, "a merger of role with person."[28] Turner's approach takes on a deterministic bent, which he initially had criticized, when he states:

> The idea of role-person merger is offered as a more behavioral complement to the subjective idea of self-conception. Careful study of the correspondence and discrepancy between self-feeling and role-person merger should enhance our understanding of the person as a social product.[29]

Careful study means an analysis of the *determinants* that *cause* the merger of individuals with their institutionalized roles. In a recent work, where he tries to develop explanatory scientific propositions, Turner offers functionality as a determinant of role differentiation.[30]

For Turner, other people and the social situation define the conditions under which the role-person merger take place. Lost is the dynamic conception of an individual who makes rather than takes roles. Turner thus offers the same determined views of roles he initially sought to avoid.

Richard Hilbert and the View of Role as an Organizing Process

Richard Hilbert believes that "role" should be treated not as a set of behaviors but as an organizing concept.[31] Rejecting Turner's reliance on norm consistency, Hilbert argues that Turner never transcends the functionalists' insistence on social order and conformity. Instead, he offers the intriguing (at least for a humanist sociology) notion that individuals don't need rule-governing behavior, that instead, role behavior is based on individuals' practical needs as they go about their everyday lives. In essence, Hilbert offers the possibility that individuals, when they need rules to prescribe, predict, or explain behavior, create them—not in relation to what Turner calls "norm consistency" but casually, when needed.[32]

According to Hilbert, this has been borne out in research by studies showing that

> actors sometimes get along fine without rules and that when they do use rules, it is *they* who require them and not the mere facticity of their behavior. In requiring them, actors sometimes use them as prescriptive guides, but in this capacity rules guide actors not by telling them exactly what to do but rather by serving notice that whatever happens may have to be rendered accountable in terms of the very rules.[33]

Hilbert's position is to view "role" as a organizing concept that is often, but not always, used by actors in social settings. "Role" in this view is not tied to the functionalists' need for order and a deterministic view of the individual but instead portrays individuals as they create many roles in terms of what they can do with these roles. Seen in this way, roles are not behavioral matrices but practical guides to behavior created by active human beings as they interpret the world around them. Individuals use roles and rules to achieve ends. Hilbert's is a commonsense model of human behavior, one in which individuals actively create their world.

Nor does the individual constantly negotiate roles, as the symbolic interactionist holds. According to Hilbert, individuals do not

> necessarily engage in continual role negotiation. If analysts insist that actors simply be continually following rules or negotiating roles even when actors do not require it, this is because analysts *require it themselves*. . . . Negotiations are not logical requirements but *members'* requirements, and in that capacity questions about who negotiates, what, when, why, and under what circumstances can only be answered by direct empirical investigation.[34]

Although Hilbert offers a dynamic conception of roles, he does not spell out the process beyond saying it is an organizing one; nor does he look at the conflicts and contractions inherent in social situations, something a humanist sociology must do.

We now turn to the post-modernists to see if they have anything to add to a dynamic theory of roles.

Post-Modern Theories of Roles

Social-psychologist Kenneth Gergen professes a belief in a dynamic self, an authentic self that defines our behavior. He sees the self as being under siege due to what he calls "social saturation."[35] Social saturation is a product of the post-modern world, a world defined by fragmentation, disintegration, malaise, meaninglessness, vagueness, and an absence of moral parameters.[36]

Social saturation furnishes the individual with a multiplicity of incoherent and unrelated images of the self. Everything that individuals once knew to be "true" about themselves is now looked at with doubt and even derision. The self has become fragmented, a compilation of a multiplicity of incoherent and disconnected relationships. These relationships pull people in various directions, forcing upon them a variety of roles which preclude "the very concept of an authentic self."[37] For Gergen, the saturated self is no self at all.

As we live out our lives in the post-modern world, relativism becomes the defining characteristic of our existence. An authentic self, a true self, is reduced to relativistic determinants. "We come to be aware that each truth about ourselves is a construction of the moment, true only for a given time and within certain relationships."[38] There is no self independent of the relationships in which the individual finds him- or herself.

Gergen sees this state of affairs producing a new pattern of self-consciousness; the individual is in a condition of *multiphrenia*, where he or she swims in an ever-changing, ever-shifting current of being.[39] Self-doubt and irrationality come to the fore. The sense of "playing a role" loses any meaning, because there is no "real self" to coordinate the role playing. Gergen sees no solution to the problem of the saturated self.

On a more optimistic note, sociologist Rose Coser believes that the changes Gergen and others have described can produce a self-reflexivity that is crucial to a humanistic orientation.[40] She holds out promise through the use of *role segmentation*—which is what Goffman called the *multiplicity of selves*. The same individual has different social roles that usually do not overlap.[41]

For Coser, individuation, the awareness of who one is in relation to others, emerges under conditions of role segmentation.[42] The multiplicity of expectations produced by modern existence forces the individual to be more reflexive. The greater choices involved produce a greater amount of reflexivity. Sameness and tradition result in the lack of a basic source of disturbance, which is also a lack of a basic source for reflection. Restricted role relationships do not allow individuals to develop. A simple set of role choices offers people a narrow definition of self.

When individuals are confronted with incompatible expectations, they are required to reflect upon them and to embark on an appropriate course of action. They must decide whether to strictly follow rules or to reinterpret or even defy them. They must weigh each decision in relation to their own purposes and the purposes of others.[43] They must role make and role remake. The result, for Coser, is a dynamic conception of self and individuals who choose what they do and what they are.

Toward a Dynamic Theory of Roles

Of the previously discussed role theorists, Linton, Goffman, and Gergen have little, if anything, to offer a dynamic theory of roles and humanist sociology. Turner, Hilbert, and Coser, however, all provide a way of salvaging a dynamic conception of role. To achieve this, emphasis should be placed on Turner's notion of role-making without the overemphasis upon norm consistency. The focus should be, as Hilbert says, on the individual and not on the norms. A dynamic theory of roles would be an organizing process but of a different sort than Hilbert envisions. It is here that Coser's views take on importance for a dynamic theory of roles. Although Gergen is right when he describes the fragmentation of the self and the resultant "social saturation," this does not necessarily preclude the existence of an authentic self. As Coser points out, the multiplicity of expectations and experiences may indeed force the individual to be more reflexive. It is out of this self-reflexivity that an authentic, dynamic self can emerge.

It is here that a humanist sociology can play a large part in the development of a dynamic theory of role. Humanist sociology tells us that human beings live out their lives in institutions, but they are not just created by these institutions, they also create them. Individuals define roles for themselves and the expectations attached to these roles, and this organizing is done in institutions which are stabilized through reification and legitimation. Because individuals create institutions and reinforce them by reification, individuals who are reflexive and critical can change these institutions. The process of role-making implies a reflexivity, a critical awareness on the part of the individual. If people can be reflexive about their roles, then they can be critical about the institutional orders within which their roles are played. If reification is bestowed by individuals, it is something that can be removed; if legitimacy can be given, it can be withdrawn. This is not an easy process, but it is not impossible. This is the lesson that a humanist sociology must teach.

It is on this potential of the individual to affect change and how the society, or those who control the society, try to control the individual that the rest of this text will concentrate. Chapter 5, in discussing social stratification, shows how inequality in American society has become accepted and is used as a means of social control. Chapters 6 and 7 focus upon the major institutional orders to show how they seek to control human behavior.

Finally, in chapter 8, a humanist challenge to the traditional sociological viewpoint that stresses the inability of individuals to effectively change their society is offered. Individuals may not have been successful in changing society, but this is not because they cannot do so, but because they have been thwarted by powerful interests who have a stake in supporting the status quo. It is in discarding the deterministic inevitability of traditional sociology and accepting the potential for autonomy within the humanist perspective that sociology can fulfill its original purpose. This is what C. Wright Mills so aptly called the use of "the sociological imagination," what is embedded in the Enlightenment view that reason could and should liberate the human spirit. In Mills' words:

> The interest of the social scientist in social structure is not due to any view that the future is structurally determined. We study the structural limits of human decision in an attempt to find points of effective intervention, in order to know what can and must be structurally changed if the role of explicit decision in history-making is to be enlarged.[44]

This is the humanist sociologist's interest in social organization—to understand how and why human beings are controlled by social structures and to discover what can be done to change this.

Summary

This chapter investigated social organization—the types of routines or patterns that develop among people through time. The two most important types of social organization are culture and society. Culture was defined as everything we think, do, and have as members of a society. Society was defined as people who share a common culture and the structures or patterns they create.

The major components of culture are: language, beliefs, values, goals, norms, folkways, mores, laws, and materiel. Society consists of statuses, roles, groups, institutions, and institutional orders. The processes of reification and legitimation give stability to institutions.

Finally, given the pivotal significance of role as a connection between the self and culture and society, we looked at role-theory, which was seen as being deterministic. Using the insights of Ralph Turner, particularly his conception of role-making as opposed to role-taking, along with Richard Hilbert's view of role as an organizing concept and Rose Coser's use of role segmentation, the possibility of a dynamic theory of roles was offered. Such a view, when incorporated into a humanist sociology, can provide the means for changing society through self-reflexivity and critical analysis.

Notes

1. Harold Garfinkel, *Studies in Ethnomethodology* (Englewood Cliffs, NJ: Prentice-Hall, 1967).

2. Ibid., p. 3.

3. This view is developed in Joseph A. Scimecca and Arnold Sherman, *Sociology: Analysis and Application* (Dubuque, IA: Kendall/Hunt, 1992).

4. Robert Bierstedt, *The Social Order*, 3d ed. (New York: McGraw-Hill, 1970), p. 123.

5. Ibid., p. 159.

6. Dorothy Lee, *Freedom and Culture* (Englewood Cliffs, NJ: Spectrum Books, 1959), p. 20.

7. Quoted in Leonore Tieffer, "The Kiss," *Human Nature*, July 1978, p. 30.

8. William Graham Sumner, *Folkways* (Boston, MA: Ginn, 1906).

9. Peter L. Berger and Thomas Luckmann, *The Social Construction of Reality* (Garden City, NY: Doubleday, 1966).

10. Charles Horton Cooley, *Social Organization* (New York: Schocken, 1962).

11. See, in particular, Peter M. Blau, *The Dynamics of Bureaucracy*, 2d rev. ed. (Chicago, IL: University of Chicago Press, 1973).

12. Bierstedt, *The Social Order*.

13. Berger and Luckmann, *Social Construction of Reality*, p. 52.

14. Ibid., p. 55.

15. Max Weber, *The Theory of Social and Economic Organization*, Talcott Parsons, ed. (New York: Oxford University Press, 1947).

16. The definitions for the political, economic, kinship, educational, and religious institutional orders are based on Hans Gerth and C. Wright Mills, *Character and Social Structure* (New York: Harbinger Books, 1964), p. 26.

17. Jacob L. Moreno, ed., *The Sociometry Reader* (Glencoe, IL: The Free Press, 1960), p. 80.

18. Ralph Linton, *The Study of Man* (New York: Appleton-Century-Crofts, 1936), pp. 113–14.

19. Orville G. Brim, "Socialization after Childhood," in Orville G. Brim and Stanton Wheeler, *Socialization after Childhood: Two Essays* (New York: Wiley, 1966), p. 3.

20. Erving Goffman, *The Presentation of Self in Everyday Life* (Garden City, NY: Doubleday, 1959).

21. Jonathan Turner, *The Structure of Sociological Theory*, 5th ed. (Belmont, CA: Wadsworth, 1991), pp. 447–71.

22. Goffman, *Presentation of Self in Everyday Life*.

23. Ibid., p. 48.

24. Ibid., p. 229.

25. See Erving Goffman, *Encounters: Two Studies in the Sociology of Interaction* (Indianapolis, IN: Bobbs-Merrill, 1961); *Behavior in Public Places: Notes on the Social Organization of Gatherings* (New York: Free Press, 1963); *Frame Analysis* (New York: Harper and Row, 1974).

26. See, in particular, Ralph H. Turner; "Role-taking: Process Versus Conformity," in Arnold Rose, ed., *Human Behavior and Social Process: An Interactionist Approach* (Boston, MA: Houghton-Mifflin, 1962), pp. 20–40, and Ralph H. Turner and Norman Shosid, "Ambiguity and Interchangeability in Role Attribution: The Effect of Alter's Response," *American Sociological Review* 41(1976): 993–1006.

27. Turner, "Role-taking," p. 23.

28. Ralph H. Turner, "The Role and the Person," *American Journal of Sociology* 84(1978):1.

29. Ibid., p. 3.

30. Ralph H. Turner and Paul Colomy, "Role Differentiation: Orienting Principles," *Advances in Group Processes* 5(1987):1 47.

31. Richard A. Hilbert, "Toward an Improved Understanding of 'Role,'" *Theory and Society* 10(1981):207–22.

32. Ibid.

33. Ibid., p. 216.

34. Ibid., p. 222.

35. Kenneth J. Gergen, *The Saturated Self* (New York: Basic Books, 1991).

36. Jean Baudrillard, *In the Shadow of the Silent Minorities* (New York: Semiotext, 1983).

37. Gergen, *Saturated Self*, p. 7.

38. Ibid., p. 16.

39. Ibid., pp. 73–74.

40. Rose L. Coser, *In Defense of Modernity* (Stanford, CA: Stanford University Press, 1991).

41. Goffman, *Encounters*.

42. Coser, *In Defense of Modernity*.

43. Ibid., p. 66.

44. C. Wright Mills, *The Sociological Imagination* (New York: Oxford University Press, 1959), p. 174.

CHAPTER 5

Social Stratification and Inequality

Although Americans like to believe they live in a middle-class society composed of individuals who are reasonably equal to one another in opportunity and prestige, even a cursory look at the United States would show that people are not regarded as equal, do not act as though they are equal to one another, and do not have equal access to what is valued. Why is there a gap between what individuals believe and the actuality of their social existence? One answer is that there is a myth of equality at work, a myth that serves as a legitimizing ideology for the stabilization of society. So long as people believe that everyone in the society has a good chance at success, at obtaining what is valued, they will not question the social and political arrangements of the society. As Benjamin DeMott put it: "Social wrong is accepted in America partly because differences in knowledge about class help to obscure it, and the key to these differences is the degree of acceptance of the myth of classlessness."[1]

The socialization process becomes a legitimizing process by emphasizing what Richard Sennett and Jonathan Cobb called the "badge of individual ability."[2] So long as people believe in the *individualistic perspective* (that they are responsible for their own success or failure), they will not try to change the status quo.

Because the self is a social creation (see chapter 3), people accept the legitimacy of the society. To question the idea that individual ability is responsible for success would mean to question an integral part of the individual's image of self. Hence, successful individuals are unwilling to question the status quo, for to do so would be to question their own self-worth.

On the other side of the coin, it is even more important for those that fail to believe that their failure is caused by their own lack of ability. If they were to think otherwise they would not accept the society as legitimate; they would not believe that society has provided equal and fair opportunity for them. Individuals who accept their lack of success (success in contemporary American society is defined in terms of valued goods and services) do not examine the social organization of society to find the root of their problem but instead are socialized to look inward. The individual's lack of success lies within. This idea is reinforced by the many popular psychology books which tell the reader how to change *his* or *her* personality, how to overcome *his* or *her* flaws, how to succeed if *he* or *she* will really try. By changing *oneself, one* eventually can overcome *one's* individual inadequacies. Should the individual feel inadequate in the changing of the self, help is at hand in the person of the therapist.

What is left unsaid is that usually the successful are successful because they were born to successful parents, that they have dominant personalities because they have been giving orders most of their lives. There is a wide gulf between the powerful and the powerless in American society, and because both refuse to acknowledge the extent and persistence of this gap, it can be said that legitimacy is well entrenched. And once such legitimacy has been established it becomes an effective justification for the manner in which power is exercised and the most *effective* argument against any attempt to change the nature of the social order. The process of legitimation, then, is a very important mechanism for the social control of individuals and is essential for an understanding of the persistence of inequality.

The hierarchy of statuses and roles enables some people to have more of what is desirable. Inequality is a social phenomenon, and social inequality is part and parcel of the structure of American society. To understand the organization of social inequality and how it affects the lives of people, it is necessary to look at the work of two giants of social science, Karl Marx and Max Weber. From their investigations of the structure of societies emerge the major questions about social inequality and social class (one's strata or group position) asked today.

Karl Marx

Marx defined social classes in relation to the production of goods. In his analysis of industrial society he named two: the *bourgeoisie,* or those who owned the means of production, and the *proletariat,* or those who worked for the owners. Originally Marx had included a third class, the landowners, whose source of income was rent. But with the growth of capitalism, this group ceased to be important in his eyes. Marx was also aware of other social groupings—artisans, merchants, intellectuals, and small land-owners—but it was only the bourgeoisie and the proletariat who were important to an understanding of the dynamics of capitalist society. To Marx, history was the story of conflict between these two classes. As he stated in *The Communist Manifesto*:

> The history of all hitherto existing society is the history of class struggles. Freeman and slave, patrician and plebian, lord and serf, guild-master and journeyman, in a word, oppressor and oppressed, stood in constant opposition to one another, carried on an uninterrupted, now hidden, now open fight, a fight that each time ended, either in revolutionary re-constitution of society at large, or in the common ruin of contending classes.[3]

Together with a permanent relationship to the means of production, individuals in a social class should exhibit a feeling of "we-ness," a togetherness that produces a psychological unity of interests. In order to constitute a class, people within the same economic strata should recognize their sameness of conditions and should feel separate from other classes. They should possess what Marx called *class consciousness*. Thus, persons in the United States from wealthy families who have gone to exclusive preparatory schools, Ivy League colleges, and professional schools should have a consciousness of their similar interests. They have been socialized to see the social world in a particular way. They have a class consciousness.

Marx never fully spelled out his view on class beyond these basic tenets. At the point where he began to elaborate upon his definition of class, in the third volume of *Capital*, the manuscript breaks off. What is left is an incomplete analysis of the dynamics of class struggle and inequality. Unfortunately, there are almost as many interpretations of Marx as there are interpreters. One of the more lucid interpretations is offered by Marxist sociologist Charles Anderson. His position is that

> inequality is not a simple phenomenon, but rather encompasses subinequalities of a diverse nature. Purchasing power, possessions, property relationships, control over production, work requirements, work security, and alienation all go into the making of inequality of economic power. These various components of inequality are closely articulated with one another and together give rise to social class divisions. These class traits tend to cluster in one direction or another so that the propertied enjoy the positive manifestations of them and the propertyless the negative manifestations.[4]

Key Marxian ideas contributing to the study of social inequality include the following: (1) economic production is the basis of social class distinctions; (2) under capitalism, the bourgeoisie own the means of production, and the proletariat work for them; and (3) the bourgeoisie dominate the proletariat through the control of ideas.

Numerous criticisms have been raised against Marx's view of class and social inequality. First, Marx predicted that after the socialist revolution the state (government) would disappear, and the proletariat would govern peacefully. Contrary to Marx's expectation, in every instance where there have been revolutions in the name of socialism, the state not only has continued to exist but has expanded.

Second, Marx assigned too much significance to class conflict, neglecting the fact that because of the process of legitimation—people accept authority as being right and proper—societies are fairly stable. While conflict does not have to be overt and can exist beneath the surface of a society, legitimation is a powerful force for social control, and Marx overlooked this.

Third, Marx contended that the gulf between the proletariat and the bourgeoisie would become wider and wider with the expansion of capitalism, and the middle class would disappear. This has not been the case in modern societies. The middle class has expanded in capitalist societies, most people have enjoyed a rise in their standard of living, and social services have been increased by governments. Finally, beginning in the late 1980s, communist and socialist countries began to restructure their economies along free-market, profit-motive lines.

Marx was accurate in that many of the conflicts in the world today are not only political and religious but also economic confrontations between the haves and have-nots. However, given the above criticisms of Marx, many sociologists have turned to the writings of Max Weber, who was one of the first social scientists to revise and expand upon Marx's thoughts on social class and social inequality.

Max Weber

Stratification, the social organization of inequality, was for Max Weber a multivariable, multidimensional phenomenon. Unlike Marx, he saw social inequality as having three aspects: *class* (income), *status* (prestige), and *party* (power).

Class

Although Weber's conceptualization of class is similar to Marx's, there is an important difference. Marx related class to the means of production, but Weber related class to power. Weber defined a class as a number of people having in common a specific causal component of their life chances, insofar as this component is (1) represented exclusively by economic interests in the possession of goods and opportunities for income and (2) represented under the conditions of the commodity or labor market.[5]

For Weber, a class was a group of people who had the same economic opportunity, lived under similar economic conditions, and possessed similar power in any given society. For example, partners in a prestigious law firm would have the same economic opportunities, live in similar conditions, and possess a similar amount of power. They would have employment with high incomes, luxurious homes, and a high level of power. At the opposite pole would be unemployed, high school dropouts living in an urban ghetto who possess no power. Weber did not emphasize, as Marx did, that the lawyers and the unemployed had to be conscious of their class position; they merely had to be an aggregate of people in similar economic positions. Class also represented life chances, or the probabilities for or against gaining what is defined as desirable in a society.

> Everything from the chance to stay alive during the first year after birth to the chance to view fine art, the chance to remain healthy and grow tall, and if sick to get well again quickly, the chance to avoid becoming a juvenile delinquent—and very crucially, the chance to complete an intermediary or higher educational grade—these are chances that are crucially influenced by one's position in the class structure of a modern society.[6]

These chances are part of being born into a particular social class.

Status

Status pertains to prestige or the ability to command deference. Class and status offer different ways in which people can be seen as unequal. For example, a drug dealer who has an annual income of $500,000, drives a Rolls-Royce, wears the finest clothes, and dines in the best restaurants would be in a high class position. But because his way of life is disapproved of by society, he would be in a low-status group with little prestige. Conversely, college professors can be in a low (economic) class position, but can command great deference. However, class and status usually overlap. Physicians are an example of one group with both high (income) class and high (prestige) status.

Status is extremely important, because along with power, it is the most directly relevant dimension of the self. A person's level of self-esteem can be high or low depending upon how the person thinks he or she is judged by others. The drug dealer would be expected to have low self-esteem, the college professor high self-esteem.

Party

When Weber used the term *party,* he meant power. A party is an organization which attempts to secure "power for its leaders in order to attain ideal or material advantages for its active members."[7] Weber was not referring to what is usually thought of as political parties but to any group that is involved in political action.

Although power is central to social ranking and social inequality in American society, it is still an elusive concept. Some distinctions therefore are useful. The most important kind of power is political power, which can be either legitimate or illegitimate.

Legitimate power is restrained power which is institutionalized and regulated by formal rules and procedures for the giving and taking of commands, positions of authority, the powers to be given to those in command, and those who will be controlled by them.[8] Illegitimate power occurs when individuals overstep the boundaries of legitimate power either by force or by manipulation.

Although Weber's writings on political power were much more brief and less well defined than were his writings on class and status, the following implication can still be drawn—class, status, and power are interrelated, and both class and status are dependent upon power.

Before examining the class structure in American society, two other aspects need to be added to Weber's three: education and occupation. These additions are important because in contemporary American society, they have become integral parts of class, status, and power.

Education

Education is defined as the amount of formal schooling one has obtained. Historically, education as it is practiced today is a relatively new phenomenon. The mass schooling that characterizes American society is a direct outgrowth of industrialism and the rise of technology. Less than two centuries ago, when most people lived in a rural environment, children learned from their parents everything they needed to know in order to take their place in society. In cities, children also learned from their parents before becoming apprentices in a trade. The wealthy hired tutors for their children before sending them to a university or abroad.

Today, mass education is a fact of life in industrial nations like the United States, and it has become impossible to learn a profession without long years in school. In America at the turn of the century, one could become a lawyer by apprenticing oneself to a lawyer and reading the law; today one must spend four years in college, then three years in law school. For the most part, the less formal schooling one receives, the less opportunity one has for achieving success. And because success usually means high class, status, and power, education becomes a basic dimension of stratification.

Occupation

Occupations are those activities which provide a livelihood or a source of income and thus are connected with *class* position. Because occupations also are a source of prestige, they are related to the status dimension. More important, occupations are a source of power, and one's occupation determines how much power one has.

How Many Social Classes?

Now we will turn to the question, How many classes are there in the United States? The answer to this question depends on who gives the answer. Sociologists generally agree that there are six classes. The six-class distinction is usually traced to Lloyd Warner and Paul Lunt's now classic division of American society into upper-upper, lower-upper, upper-middle, lower-middle, upper-lower, and lower-lower classes.[9]

Warner and Lunt's Conception of Social Class

While there is some disagreement over the percentages of the population in each class, those who accept the six-classes division usually agree that about 1 percent of the population falls into the upper-upper class. This includes families who have accumulated wealth over two or more generations. The next division—lower-upper—is made up of about the same percentage of people, but the wealth of this group has been more recently acquired.

The third category—upper-middle—contains about 10 percent of the population and is comprised of successful business and professional people. The lower-middle class has about 30 percent of the population and is composed mostly of white-collar workers.

The largest class is the upper-lower, which represents about 33 percent of the population and is made up of blue-collar workers. Approximately 25 percent of the remaining population is in the lower-lower class and includes unskilled workers and the unemployed.

Gilbert and Kahl's Conception of Social Class

A more recent conception of the composition of the six social classes has been offered by Gilbert and Kahl.[10] They see the class structure in terms of nine variables: occupation, income, wealth, personal prestige, association, socialization, power, class consciousness, and mobility. With the exception of power, which is multidimensional, they believe that these variables can be measured by distinct and separate operations, and each can be used to stratify various populations. These nine variables, although not sufficiently linked to form a strict theory of stratification, still offer a useful conceptual scheme that describes the class system.

Gilbert and Kahl describe the six classes and their percentages as follows:

1. *Capitalist Class.* A class whose income is derived largely from returns on investments and assets. It consists of about 1 percent of the population.
2. *Upper-Middle Class.* Examples are university-trained professionals and managers. A few from this class will rise to the top of the corporate hierarchy and ascend into the capitalist class. It consists of about 14 percent of the population.
3. *Middle Class.* They follow the orders of the upper-middle class. They receive sufficient income to make a good living and enjoy a comfortable style of life. There is some advancement from this stratum to the upper-middle class. Most are white-collar workers, but some are blue collar. This class consists of about 25 percent of the population.
4. *Working Class.* These people are less skilled than individuals in the middle class. They usually work at routine, closely supervised manual and clerical jobs. Although they receive a steady income, they have relatively little chance of advancing in the stratification hierarchy, because they lack the educational credentials. Seniority is more important than promotion. This class consists of about 30 percent of the population.
5. *Working-Poor Class.* People employed in such low-skill jobs as laborers and service workers make up this class. Their income is well below the median, and they cannot depend on steady employment, given the nature of their jobs. Rather than hoping for advancement in the stratification system, they are at risk of dropping down to the class below them. This class consists of about 20 percent of the population.
6. *Underclass.* Members of this class have little or no participation in the labor force. Employment is erratic or part-time. Many lack skills and education and have difficulty finding full-time permanent jobs. Others with skills cannot find full-time employment because of structural dislocation (plant closings, etc.). They depend upon gov-

ernment programs for assistance. This class consists of about 10 percent of the population.[11]

A Seven Social Class Conception

My view is that there are seven social classes in American society[12]: upper-upper (1 percent), lower-upper (1 percent), upper-middle (7-8 percent), middle-middle (9-10 percent), lower-middle (25 percent), working (30 percent), and lower class (25 percent). The addition of the category "middle-middle" to the traditional divisions is based on the growth of bureaucracy and large corporations (discussed in chapter 6) and the subsequent need for middle-level executives in American society.

Now we will look more closely at the seven class system in the United States as I see it.

The Upper-Upper Class

The upper-upper class is composed of older families whose wealth was accumulated after the Civil War. Their fortunes were made in the newer industries of banking, chemicals, steel, railroads, shipping, oil, and automobiles. The DuPont, Harriman, Rockefeller, Mannville, and Ford families would be typical examples of this class, which emphasizes "old money." The upper-upper class sends their children to the finest preparatory and finishing schools and to Ivy League and Seven Sister colleges, and expects them to carry on in the tradition of their social class. Membership in this class is based on lineage: a male must be born into it, though a female may marry into it. A case in point is the late president of the United States John F. Kennedy. Although the son of one of the wealthiest men in the country and elected to its highest office (which made him one of the most, if not the most, powerful man in the world), John Kennedy was not considered to be a member of the upper-upper class. His father, Joseph Kennedy, had made the family fortune and therefore the Kennedy's wealth was not old enough. Jacqueline Bouvier Kennedy Onassis, on the other hand, although her family's wealth was nowhere near the Kennedy's, came from an old enough family to be considered upper-upper class. The Kennedy children, John and Caroline, because their grandfather made the family fortune and because they have been to elite schools, are members of the upper-upper class. Upper-upper class persons may work at a profession and serve on numerous boards of directors in banking, business, and education.

The Lower-Upper Class

Membership in this class is based primarily on newly acquired wealth. The older families of the upper-upper class regard the lower-upper members as "upstarts." They are the *"nouveaux riches,"* and only when their children and grandchildren attend the finest schools and universities and they are socialized into what Joseph Kahl characterized "as graceful living"[13] will they be admitted into the upper-upper class—provided, of course, that they hold on to their wealth. Examples of members of the lower-upper class would include such family names as Hunt, Murchison, Perot, Getty, and Walton.

The Upper-Middle Class

The typical upper-middle class person is a successful professional (doctor, lawyer, accountant) or top executive in a large corporation. Upper-middle class people are very much achievement oriented and try to instill this in their children. Education, in particular, is stressed, because it was success in education that helped them gain their social class position.

Even though the Women's Movement has had its greatest success in the middle class, its support in the upper-middle stratum is more mixed than in the middle-middle and lower-middle classes. The husband is still the major breadwinner in the upper-middle class, and the family's status revolves around his career. Although the wife may also be a professional, she is expected to neglect her own career or pick up and move with her husband should he get a better job opportunity. Very rarely is this reciprocated by the husband. That the Republican ticket of Bush and Quayle could in 1992 make an issue of Hillary Clinton's having a career of her own, implying a lack of "family values" on the part of the Clintons, attests to the entrenchment of this view.

The Middle-Middle Class

The middle-middle class is composed of college-educated, white-collar workers who are mainly in middle-level positions in large corporations or government. They differ from their upper-middle class counterparts in that they lack the occupational autonomy of a doctor or a lawyer; they do not have as much freedom on the job and cannot set their own working hours. Their income does not approach that of the upper-middle class professional. Usually both husband and wife work, frequently at the same occupational level. Indeed, it is the two paychecks that enable them to pay the high mortgage on their suburban home and meet the two car payments that are necessitated by their two careers. Like the upper-middle class person, the middle-middle class members place a high premium on education, hoping that educational success will enable their children to surpass them and ascend into the upper-middle class.

The Lower-Middle Class

Lower-middle class persons make less money and have completed fewer years of schooling than their middle-middle class counterparts. Occupational examples are data processors, bank tellers, and salespeople. Their salaries are much lower than those of the middle-middle class; they cannot look forward to living comfortably.

The lower-middle class is extremely status conscious; respectability is a constant and pressing concern. This emphasis upon respectability often takes on a strong moralistic tone. Status consciousness is seen as a perennial problem, given that those below them on the stratification ladder (in particular, highly skilled blue-collar workers) often have higher incomes. Status must therefore be stressed in order to enable this stratum to preserve its tenuous middle-class identity. Thus, many lower-middle class families will make many sacrifices in order to move to the suburbs, believing that such a move will enhance their status. Suburban schools are also seen as providing a better education for their children, thereby enabling them to ascend higher in the stratification system.

The Upper-Lower or Working Class

Upper-lower or working-class people are in a difficult position. Often their weekly income is about equal to or less than their expenses. This creates difficulties, especially when unplanned bills arrive. The negative cash flow plus the low self-esteem of people in this social class create additional problems. Members of this class tend to possess a high degree of personal insecurity, which often results in a tenuous self-image. They feel ill at ease in the company of those of higher status, given their lack of education and what they perceive as inadequate social graces, and feel threatened by those of lower status, who are seen as receiving benefits without earning them. Blacks, in particular, are singled out by white members of the working class as gaining from affirmative action programs at the expense of whites. White working-class members have had problems accepting the original intention of affirmative action—the assurance that minorities are not discriminated against when considered for promotion and/or hiring.

Although working-class parents have high expectations for their children, they lack the power or resources to help their children succeed. They have high expectatations because they share the belief that hard work and education will make it possible for their children to rise in the social structure. They hold this belief despite the evidence of their daily lives. Only the most intelligent or talented children of this stratum achieve success. Working-class parents are forced to rationalize their children's lack of success by pointing out that "they're doing all right" without having been to college or that they are holding down a steady job. In the end, because they themselves have little control over their own lives, they cannot muster enough resources to help their children achieve a better life than their own.

The Lower-Lower Class

Lower-lower class individuals are at the very bottom of the class structure. They are characterized by low-prestige semi-skilled occupations (at best) and unemployment (at worst), low or no salaries, little education, and little if any power to control their own or their children's destinies. They are trapped in a cycle of poverty from which few escape. Generation after generation, the lower-lower class populates the unskilled and unemployed labor force. Remaining in poverty and lacking dignity, they are discriminated against and stereotyped as uninterested in helping themselves. Given their powerlessness, they can do little to affect their position in American society.

The Extent of Inequality

Now that we have an idea of the composition and characteristics of the various social classes, we can turn to the major questions that must be asked about inequality: (1) What is the extent of inequality? and (2) Is inequality inevitable? The two most important forms of inequality are wealth and occupational mobility.

Inequality in Wealth

An important indicator of economic inequality is "total money income" or the sum one receives from earnings. Such an index is readily available from the U.S. Census Bureau and offers a constant base against which individuals can be compared. Even more important is the more complex index of a family's "total wealth," which includes all assets—homes, automobiles, savings, investments, and so forth—minus debts.

Census Bureau figures show that the poorest one-fifth of the nation owns 4.5 percent of the national wealth, while the richest one-fifth owns 44.2 percent.[14] Furthermore, 2 percent of the population owns 45 percent of all corporate stock.[15] Economist Lester Thurow also tells us that a few hundred families control 40 percent of the fixed nonresidential capital of the United States.[16]

The Poverty Level

Since 1969 the official poverty level used by the U.S. government has been determined by a scale developed in 1965 by economist Mollie

Orshansky for the Social Security Administration. Orshansky defined the poverty level in terms of minimum income that could support families of various sizes. The poverty line is calculated "on the amount needed by families of different size and type to purchase a nutritionally adequate diet on the assumption that no more than a third of the family income is used for food."[17] In 1993, the poverty line for a family of four was $14,335, $11,786 for a family of three, and $7,143 for a single individual. Obviously, this is a rough approximation of poverty, for many other factors must be considered. For example, government subsidies are not included. Also, current costs of housing and food show that it costs more to live in an urban than a rural area. And the assumption that one-third of a family's income is spent on food is probably not valid. However, despite the tendency of the government calculations to posit less poverty than there actually is, in 1993 there were 36.9 million people (14.5 percent) in the United States whose income was below the poverty line. This figure represented an increase of 1.2 million over the previous year.[18] Furthermore, 21.9 percent of all children live in poverty,[19] and it has been estimated that 45 percent of all black children live in poverty.[20] It should be noted, too, that while almost half of black children live in poverty, roughly two-thirds of the poor in the United States are white.

Comparative data also show the United States lags behind other industrial nations in dealing with poverty. An Urban Institute study found that a higher proportion of children in the United States live in poverty than children in seven other industrial democracies. This was the case even though family incomes in the United States were higher than in most of the seven other nations.[21]

Perhaps nowhere does this inequality prove to be more harmful than in the relationship of poverty and health. In general, the higher one's class, the longer one's life expectancy. "This relationship has been found in different countries, at different times, and with different study procedures."[22] Blacks, who are disproportionately represented in the lower classes, have a shorter life expectancy than whites. The average age expectancy at birth for a black female is 75.2 years; for a white female, it is 79.4 years. A black male baby has a life expectancy of 67.0 years in contrast to 72.7 for a white male baby.[23]

In health interviews, those with low incomes report that they are in poorer health than those with high incomes. The U.S. Public Health Service found that low-income people (under $5,000 a year) were likely to have numerous "restricted activity days," greater health-related work-time losses, and more frequent treatment for severe conditions that required medical care.[24] Nor is the relationship between social class and health restricted to physical health. It has been shown that the prevalence of mental illness varies inversely with social class.[25] And finally, it needs to be noted that the U.S. Census Bureau reported that in 1992, 37.4 million people had no health insurance.[26]

What exists, then, is a very real gap between the "haves" and the "have-nots" in this country. Indeed, two economists, William Goldsmith and Edward Blakely, in a recent review of governmental statistics on poverty, came to the conclusion that the poor in America were living in a "separate society" when compared to the rest of the nation. "For nearly fifty years the nation committed itself to reduce poverty, equalize resource distribution and augment the middle class. These improvements, incomplete though they were, have been reversed. Economic and political forces no longer combat poverty—they generate poverty."[27]

Furthermore, the gap between those at the top and those at the bottom has continued to increase during the past quarter of a century. In 1992, the most affluent one-fifth of all families had incomes that averaged 8.4 times the poverty level as opposed to 6 times the poverty level in 1967. In contrast, the least affluent one-fifth of families had incomes averaging 91 percent of the poverty level in 1992, as opposed to 97 percent in 1967.[28]

An important way in which this separation, this gap between the rich and the poor, has been perpetuated is through the tax structure.

The Tax Structure

The tax structure is a principal means through which the rich in American society hold on to their wealth. There are two kinds of taxes: *regressive* and *progressive*. Regressive taxes take a steadily smaller proportion from an individual as income increases. Progressive taxes do just the opposite, taking a steadily increasing proportion of income as it rises. Sales taxes, property taxes, and Social Security taxes are examples of regressive taxes. The federal income tax is most often used as the prime example of a progressive tax in the United States. However, upon close inspection it turns out that the federal income tax is not all that progressive. For example, during the 1980s, when tax reforms were initiated, an unmistakable trend toward more inequality was exacerbated by the legislated changes in the federal income tax structure. While the top 1 percent paid 23.2 percent less in federal taxes during the years 1977 to 1990, the lowest 20 percent paid 2.6 percent more.[29] Much of this unfairness can be explained by the fact that the higher income categories include several income categories other than wages, whereas the lower categories are dominated by wage earnings. Wages generally provide less opportunity for reducing or avoiding taxes. Perhaps the greatest unfairness lies in the manner in which the federal tax laws favor large corporations, often at the expense of the worker. Corporations can make money by closing plants and factories and writing off the estimated value of their closing against profits made in other units of the corporation. Union Carbide once earned $620 million in tax savings by closing chemical plants. United Technologies received $424 million for closing down a computer-equipment subsidiary. TRW wrote off $142 million in tax liabilities by halting domestic production of certain aircraft components.[30] Each of these tax bonanzas and others like them had enormous consequences for the work force throughout the nation and led in the early 1990s to the largest unemployment rates in over a decade.

Tax laws are written to benefit the rich. They are ingeniously complex, and the public is kept in ignorance as to the real beneficiaries of these laws. The middle class does receive some benefit from certain tax laws, but these benefits are hardly comparable to what the wealthy receive. The allowable deduction on mortgages that middle-class home owners are permitted to take provides just enough benefit to keep them from raising any meaningful questions about the tax laws. Thus, an unfair tax structure is perpetuated. Income and wealth inequality have become a fact of life in the United States. We will now look at occupational mobility to see if conditions are more equal there.

Occupational Mobility

Occupational mobility, or whether people are able to rise in society, has often been used to gauge inequality. Is there truly social mobility based on occupation? How much mobility has there been? What is the nature of the mobility that has occurred? Is American society as open as most people believe?

American sociologists, for the most part, have equated social class with occupation and have looked at social mobility as occupational mobility. The reason for this is that occupation is a very important part of an individual's life. Peter Blau and Otis Duncan, two sociologists who have studied this aspect of inequality, claim:

> The occupational structure in modern industrial society not only constitutes an important foundation for the main dimensions of social stratification but also serves as the connecting link between different institutes and spheres of social life, and therein lies its greatest significance. The hierarchy of prestige strata and the hierarchy of economic classes have their roots in the occupational structure; so does the hierarchy of political power and authority, for political authority in modern society is largely exercised as a full-time occupation. It is the occupational structure that manifests the allocation of manpower to various institutional spheres, and it is the flow of movements among occupational groups that reflects the adjustment of the demand for diverse services and the supply of qualified

manpower. The occupational structure is also the link between the economy and the family, through which the economy affects the family's status and the family supplies manpower to the economy.[31]

Here, positivist sociologists have substituted something that is easily measured (occupation) for something that is not (social class). The equation of occupation with social class is suspect and should be examined at length.

Early Studies of Social Mobility

One of the first studies to deal with the question of occupational mobility was carried out by Natalie Rogoff in 1953. Looking at intergenerational (father's occupation compared to son's occupation) mobility in Indianapolis, she found that there had not been any major changes in the extent of intergenerational mobility from 1910 to 1940. In both periods the occupations of the sons were basically the same as those of their fathers.[32] A follow-up study in 1970 by Tully, Jackson, and Curtis found that occupational mobility had changed very little since the Rogoff study.[33]

Because Rogoff and Tully et al. studied only one city, other sociologists raised criticisms concerning the generalizability of their findings. It was only with Blau and Duncan's publication of their now famous *The American Occupational Structure* in 1967 that sociologists looked at national data. Using a sample of over 20,000 males (women were not in the work force in significant-enough numbers to be included in the sample), Blau and Duncan traced the interdependence of what they considered to be four determinants of occupational achievement: father's occupation, father's education, son's education, and son's first occupational position.[34]

Blau and Duncan concluded that there was no evidence of lack of mobility in the occupational structure, and that upward mobility had increased slightly since World War II. They also suggested that the role of education was becoming more and more important to occupational achievement, which in turn produced more mobility.[35] In a replication of the Blau and Duncan study, David Featherman found that Americans "enjoy at least as much opportunity for socioeconomic mobility as in early periods of this century."[36]

While the Blau and Duncan study still is an important work about mobility, several questions can be raised. First, though Blau and Duncan claim that there is mobility in the occupational structure, a close look at their data reveals that there was more mobility *within* white-collar occupations and *within* blue-collar occupations than *between* these two strata.

Second, the intergenerational model assumes that the occupational categories found in the father's generation are similar in the son's. Given the rapid industrialization that characterized this nation during the time of Blau and Duncan's study, this is not the case.[37] Furthermore, some of the fathers in the earlier generation did not produce sons for the labor force, some of the fathers continued to be in the labor force, and some of the sons may have been the children of an immigrant father who did not participate in the labor force.[38]

Third, Blau and Duncan use crude categories for analyzing mobility and "do not take into account the high probability that the laborer's mobile son enters a lower-status profession or business career than the professional's son, for example, possession of a teaching certificate as opposed to a Ph.D., an assistant manager of a small business as opposed to an executive in a large company."[39]

Fourth, because of their sample, Blau and Duncan tell us nothing about women, a group that has had a profound impact on the labor market in the past several decades. More recent studies that include women find that women have more downward mobility than do men.[40] Fifth, in the United States one's social status depends on income and education as well as occupation. Taking income and education into consideration, Christopher Jencks found that there is less movement up and down the social ladder than up and down the occupational ladder.[41]

Finally, even if one accepts their quantitative methods of analysis as the best available measure of mobility (which humanist sociologists do not), Blau and Duncan claim only to explain 25 percent of the relationship between father's occupation, father's education, son's education, and son's first job.

In short, even when sociologists use a not very reliable measure of social class or occupation, the evidence does not support the belief that America is a highly mobile, open-class society. Inequality in wealth and occupational mobility seems to be a fact of life. This leads us to the next question about inequality: Is it inevitable?

Is Inequality Inevitable?

The best-known rationalization for the persistence of inequality has been offered by sociologists Kingsley Davis and Wilbert Moore.[42] Using, a structural-functional argument, they held that inequality is inevitable because it is necessary; it serves a purpose or function for the survival of a given society. Accordingly, a critical problem for society is the motivation of individuals to occupy the more important statuses or positions in society. Society must provide inducement to perform its most important tasks, and social inequality is "an unconsciously evolved device by which societies ensure that the most important positions are conscientiously filled by the most qualified persons."[43] Therefore, every society has tasks that are differentially important to its survival, and because of this, every society is stratified.

Davis and Moore listed two criteria that determine the rewards that accrue to given positions: (1) the importance of the task, and (2) the scarcity of personnel capable of performing the task, or the amount of training required. These conditions together, determine the rank of a given occupation in a stratified society.[44] Consequently, "a position does not bring power and privilege because it draws a high income. Rather it draws a high income because it is functionally important and the available personnel is for one reason or another scarce."[45]

Davis and Moore have been criticized for not dealing with power differentials, which is a somewhat unfair criticism. They did imply that a third and more radical factor is also involved in determining an individual's (as opposed to a position's) rank as reward—economic power or control over resources. They recognized that having a great deal of money can give one a real advantage in getting a higher position, and they were aware that power and prestige are based on ownership.

However, Davis and Moore did not develop this point and concluded that economic considerations take a secondary place to functional importance, talent, and training. It is here that their thesis is open to question. The most important or the most prestigious positions are *not* always filled by the most qualified persons. Take, for example, the position of Supreme Court justice, which is not only a position of power but also continuously ranked by national samples as the most prestigious occupation in the United States. Although there have been many outstanding legal minds who have served on the Supreme Court, justices are usually chosen by the president of the United States on political and ideological grounds. Nowhere was this more obvious than with former President George Bush's appointment of Clarence Thomas to the Supreme Court. Bush claimed that Thomas was the "most qualified person in the country" for the position, that his appointment had nothing to do with the fact that Thomas was a black conservative.

And even if it is posited for argument's sake that the most qualified persons do enter the most important positions, another question should be considered: What are the criteria used to evaluate importance? Is a tax accountant who is paid $500,000 a year to keep large corporations from paying taxes more beneficial to society than a lumberjack who is paid $50,000 a year to keep the accountant supplied with paper for his or her work? The reason for paying the accountant a larger salary is power, not social need.

How much reward is necessary for society to insure that individuals will become high-ranking administrators and executives, physicians, scientists, professors, lawyers, and so forth? At what level of reward would occupations go begging for want of qualified persons? Would medical or law or business schools close their doors because of a lack of qualified applicants if doctors or lawyers or corporate executives were limited to a salary of $250,000 per year, a figure that would still place them among the top one-half of 1 percent of the population?

In short, what Davis and Moore have done is create an ideal situation in an abstract society, a picture of inequality that has little to do with reality. They do not show that inequality is inevitable; instead, they offer a rationalization for the individualistic perspective.

America is not the land of opportunity that it is professed to be in our schools and in the media. The United States as a classless society is a myth. What's more, social classes have been with us at least since the Revolutionary War and most likely before that. As Charles Hurst points out:

> The studies that have been produced of wealth in the early United States consistently point to the fact that wealth inequality was a clear and constant condition during that period. This was especially true for the period between the Revolutionary and Civil Wars, a time in which inequality was on the rise.[46]

Historian James Sturm, in his study of cities in the New England, Middle Atlantic, Southern, and Midwest states, found increasing inequality in estate wealth from 1800 to 1850.[47] Another study by Edward Pessen found that in Brooklyn, New York, in 1810, 1 percent of the population held 22 percent of the wealth, increasing to 42 percent by 1840.[48] Figures for Boston and New York show the same trend. In 1820, the top 1 percent of Bostonians held 16 percent of the wealth, 37 percent by 1848. In New York, the figures were 29 percent in 1828 and 40 percent by 1845.[49] Finally, Lee Soltow found that from 1850 to 1870 wealth inequality still remained high, so much so that in these two decades "there very definitely was an elite upper group in America in terms of control of economic resources."[50]

The evidence is quite clear for the historical persistence of the inequality that existed in the eighteenth and nineteenth centuries. It is myths that historically there were very few rich or poor people in America, that "most of the rich men were formerly poor," and that when wealth was accumulated it was "insecure" and circulated with "inconceivable rapidity."[51] Or that in American society everyone can succeed solely on the basis of ability and motivation. It is more accurate to say that a small number of families made their wealth early and passed on their wealth and the power that goes with it from one generation to the next.

That they have been able to do so is a function of the use of power and social control. The lower classes are socialized to accept the structure as is. Since 1959, Melvin Kohn and his associates at the National Institute of Mental Health have studied how values are transmitted from generation to generation among different social classes and how these values are used as a means of social control.[52] Looking at such values as obedience, neatness, cleanliness, consideration, curiosity, self-control, happiness, and honesty, Kohn noted clear distinctions between the social classes. Kohn found that middle-class parents stressed self-control, curiosity, and consideration, which cultivate capacities for self-direction and empathic understanding in their children; working-class parents stressed neatness, obedience, and good manners, and only cultivated capacities for behavioral conformity.

Kohn called the two types of value orientations "self-direction" and "conformity." At the higher levels of the class structure, parents valued self-direction for their children and devalued conformity to external rules. The opposite held for those lower down on the social stratification ladder.

Kohn interpreted these findings to mean that the generalized value orientation develops out of experiences in the occupational world. People who are professionals or who hold management jobs need to make independent judgments more than do supervised nonprofessional workers. This was found to hold not only in the United States but in other countries as well. All of these findings led Kohn to conclude that authoritarian attitudes stressing conformity "to the dictates of authority and intolerance of nonconformity" were more frequent at the lower levels of the social class scale.[53]

Kohn's findings are important for our understanding of the social class system as a whole. If people at the top teach their children to value self-direction while those at the bottom teach their children to value conformity to authority, then what we have is a system that prepares people to remain in the same social class as their parents. Furthermore, even if the lower classes could teach their children different value systems, it is probable that discrimination and life experiences in general would force them back into the old values, with a resultant bitterness. In short, the sociological perspective holds that the values people have reflect their social reality. In order to change, it is not enough to change attitudes and values; it is necessary to change the structure of the experiences that produced those attitudes and values in the first place. The American Dream (that one merely has to work hard to achieve success) is not being realized by those in the lower social classes who are condemned by structural conditions to a persistent cycle of poverty and inequality.

In chapters 6 and 7 we see how the institutional orders further reinforce social inequality and social control. This leads us to the most important question a humanist sociologist can ask: How can a system that is unfair, that produces inequality, be changed? Chapter 8 attempts to provide an answer to this question.

Summary

The topic of this chapter was social inequality. The classical works of Karl Marx and Max Weber on social inequality and social class were introduced, because these two great thinkers provided the framework for the questions that are being asked about inequality today.

Social class was seen to be a multidimensional term. To Weber's class, status, and power, the dimensions of education and occupation were added. Seven classes based on these five dimensions were then posited: upper-upper, lower-upper, upper-middle, middle-middle, lower-middle, upper-lower, and lower-lower. Each of the seven social classes was then characterized.

Two major questions concerning inequality were raised: (1) How much inequality is there? and (2) Is inequality inevitable? The answer to the first question is "a great deal," and the answer to the second question is that inequality is not inevitable. Instead, it is part of a structured system of inequality, a system in which, as Melvin Kohn and his associates have shown, individuals are socialized to accept a value system that reinforces social control.

Notes

1. Benjamin DeMott, *The Imperial Middle: Why Americans Can't Think Straight about Class* (New York: Morrow, 1990), pp. 10–11.

2. Richard Sennett and Jonathan Cobb, *The Hidden Injuries of Class* (New York: Vintage Books, 1973), p. 58.

3. Karl Marx, *The Communist Manifesto*, J. Katz, ed. (New York: Washington Square Press, 1964), pp. 57–58.

4. Charles H. Anderson, *The Political Economy of Social Class* (Englewood Cliffs, NJ: Prentice-Hall, 1974), p. 78.

5. H. H. Gerth and C. Wright Mills, eds. and trans., *From Max Weber: Essays in Sociology* (New York: Oxford University Press, 1946), p. 181.

6. Hans Gerth and C. Wright Mills, *Character and Social Structure* (New York: Harbinger Books, 1964), p. 313.

7. Max Weber, *Economy and Society*, Gunther Roth and Clauss Wittich, eds. (New York: Bedminster Press, 1968), p. 284.

8. Beth Ensminger Van Fossen, *The Structure of Social Inequality* (Boston, MA: Little, Brown, 1979), p. 139.

9. W. Lloyd Warner and Paul S. Lunt, *The Social Life of a Modern Community* (New Haven, CT: Yale University Press, 1941).

10. Dennis Gilbert and Joseph Kahl, *The American Class Structure* (Homewood, IL: Dorsey, 1987).

11. Ibid., pp. 345–46.

12. The following descriptions of the seven social classes owe a great deal to Joseph Bensman and Arthur J. Vidich, *The New American Society: The Revolution of the Middle Class* (Chicago, IL: Quadrangle Books, 1971). For a fuller explication, see Joseph A. Scimecca and Arnold K. Sherman, *Sociology: Analysis and Application* (Dubuque, IA: Kendall/Hunt, 1992).

13. Joseph A. Kahl, *The American Class Structure* (New York: Rinehart, 1957), p. 187.

14. U.S. Dept. of Commerce, Bureau of the Census, *Current Population Reports* (Washington, DC: U.S. Government Printing Office, 1992), pp. 60–180.

15. Board of Governors, Federal Reserve System. *Federal Research Bulletin*, Jan. (Washington, DC: U.S. Government Printing Office, 1992).

16. Lester Thurow, *The Zero-Sum Society* (New York: Basic Books, 1980).

17. Mollie Orshansky, "How Poverty Is Measured," *Social Security Bulletin* 22 (Feb. 1969):37–41.

18. Guy Gugliotta, "Number of Poor Americans Rises for 3rd Year," *Washington Post*, Oct. 5, 1993, p. A6; Robert Pear, "Poverty in U.S. Grew Faster Than Population Last Year," *New York Times*, Oct. 5, 1993, p. A20.

19. Ibid.

20. Gerald Jaynes and Robin Williams, *A Common Destiny* (Washington, DC: National Academy Press, 1989).

21. John L. Palmer, Timothy Smeeding, and Barbara Boyle Torrey, eds., *The Vulnerable.* Washington, DC: Urban Institute, 1988).

22. Judau Matras, *Social Inequality, Stratification and Mobility* (Englewood Cliffs, NJ: Prentice-Hall, 1975).

23. U.S. Dept. of Commerce, Bureau of the Census, *Current Population Reports* (Washington, DC: U.S. Government Printing Office, 1992).

24. U.S. Bureau of the Census, *Statistical Abstract of the United States* (Washington, DC: U. S. Government Printing Office, 1977, 1985).

25. See August Hollingshead and Frederick Redlich, *Social Class and Mental Illness: A Community Study* (New York: Wiley, 1958), and Leo Srole, et al., *Mental Health in the Metropolis: The Midtown Manhattan Study* (New York: McGraw-Hill, 1962).

26. Gugliotta, "Number of Poor Americans Rises for 3rd Year," p. A6; Pear, "Poverty in U.S. Grew Faster Than Population Last Year," p. A20.

27. William W. Goldsmith and Edward J. Blakely, *Separate Societies; Poverty and Inequality in U.S. Cities* (Philadelphia, PA: Temple University Press, 1992), p. 1.

28. Pear, "Poverty in U.S. Grew Faster Than Population Last Year," p. A20.

29. U.S. House of Representatives Ways and Means Committee Report, 1990.

30. Goldsmith and Blakely, *Separate Societies*, pp. 91–92.

31. Peter M. Blau and Otis Dudley Duncan, *The American Occupational Structure* (New York: Free Press, 1978), pp. 6–7.

32. Natalie Rogoff, *Recent Trends in Occupational Mobility* (Glencoe, IL: Free Press, 1953).

33. J. C. Tully, E. F. Jackson, and R. F. Curtis, "Trends in Occupational Mobility in Indianapolis," *Social Forces* 49(1970);186–200.

34. Blau and Duncan, *American Occupational Structure*.

35. Ibid., pp. 402–42.

36. David L. Featherman, *"Has Opportunity Declined in America?"* Institute for Research on Poverty Discussion Paper, no. 437–77, University of Wisconsin–Madison, 1977, p. 15.

37. To be fair, Duncan is aware of this criticism, although he does not believe it detracts from his analysis. See Otis Dudley Duncan, "Methodological Issues in the Analysis of Social Mobility," in Neil J. Smelser and Seymour Martin Lipset, eds., *Social Structure and Mobility in Economic Development* (Chicago, IL: Aldine, 1966), pp. 51–97.

38. Natalie Rogoff Ramsy, "Changes in Rates and Forms of Mobility," in Smelser and Lipset, eds., *Social Structure and Mobility in Economic Development*, p. 215.

39. Charles H. Anderson, *Toward a New Sociology*, rev. ed. (Homewood, IL: Dorsey, 1974), p. 136.

40. See Geoff Payne and Pamela Abbott, eds., *The Social Mobility of Women: Beyond Male Mobility Models* (London: Farmer Press, 1990).

41. Christopher Jencks, "What Is the True Rate of Social Mobility?" in Ronald L. Breiger, ed., *Social Mobility and Social Structure* (New York: Cambridge University Press, 1990), pp. 103–30.

42. Kingsley Davis and Wilbert E. Moore, "Some Principles of Stratification," *American Sociological Review* 10(1945):243.

43. Ibid., pp. 243–44.

44. Ibid., pp. 246–47.

45. Ibid., p. 247.

46. Charles E. Hurst, *The Anatomy of Social Inequality* (St. Louis, MO: Mosby, 1979), p. 22.

47. James Lester Sturm, *Investing in the United States, 1798–1893* (New York: Arno Press, 1977).

48. Edward Pessen, *Riches, Class, and Power before the Civil War* Lexington, MA: Heath, 1973, p. 36.

49. Ibid., pp. 33–34.

50. Lee Soltow, *Men and Wealth in the United States* (New Haven, CT: Yale University Press, 1975), p. 180.

51. Quotations from Tocqueville, cited in Edward Pessen, "The Egalitarian Myth and the American Social Reality: Wealth, Mobility, and Equality in the 'Era of the Common Man,'" *American Historical Review* 76(1971):902, 1004, 1015.

52. See, in particular, Melvin Kohn, *Class and Conformity* (Homewood, IL: Dorsey, 1959); Melvin Kohn, "Social Class and the Exercise of Parental Authority," *American Sociological Review* 24(1959):352–59; Melvin Kohn, "So-

cial Class and Parental Values: Another Confirmation of the Relationship,"
American Sociological Review 61(1976):538–65,568; and Melvin Kohn and Carmi
Schooler, "Job Conditions and Personality: A Longitudinal Assessment of Their
Reciprocal Effects," *American Journal of Sociology* 87(1982):1257–86.

    53. Kohn, *Class and Conformity*, p. 79.

CHAPTER 6

The Dominant Institutional Orders: Economic and Political

Power is *the most important variable* for understanding social inequality and social control. In modern industrial societies, power is located in institutional orders. In American society it is found in the economic and political orders (the corporate economy and the government). These two orders are dominant institutional orders which manage the others; they set the values that shape a society and its reality for its members. Subordinate institutional orders—the family, religion, education, and the mass media—legitimate the decisions made by those at the top of the dominant institutional orders. The economic and political orders dominate American society, and the remaining four socialize individuals to accept what the main two want.

This is a modification of C. Wright Mills' famous "power elite" thesis. Mills believed that a power elite, comprised of those who held the top positions in the economic, political, and military institutional orders, ruled America, or

> those political, economic, and military circles which as an intricate set of overlapping cliques shape decisions having at least national consequences. Insofar as national events are decided, the power elite are those who decide them.[1]

Whether the military order was as dominant as Mills held is debatable. Less debatable is that the military is not on the same level as the corporate economy and the government as a locus of power. It is therefore treated in this work as a part of the state or the political institutional order.

Within each of the dominant orders, the typical institutional unit has become highly centralized due to the rapid development of technology in this century. Here Mills' description seems to be on the mark.

> The economy—once a great scatter of small productive units in autonomous balance—has become dominanted by two or three hundred giant corporations, administratively and politically integrated, which together hold the keys to economic decisions.
>
> The political order, once a decentralized set of several dozen states with a weak spinal cord, has become a centralized, executive establishment which has taken up into itself many powers previously scattered, and now enters into each and every cranny of the social structure. . . .
>
> In each of these institutional areas, the means of power at the disposal of decision makers have increased enormously; their central executive powers have been enhanced; within each of them modern administrative routines have been elaborated and tightened up.[2]

The major institutional orders have become bureaucratized, as has nearly every aspect of American society.

Bureaucracy and the Structure of American Society

American society today is dominated by large-scale bureaucracies. The most influential framework within which an analysis of bureaucracy's role in industrialized society can be discussed was developed by Max Weber in the early twentieth century. To Weber, the principal characteristic of bureaucracy was its clearly specified areas of jurisdiction, which are defined by explicit rules. These rules are enforced by those who occupy positions within the bureaucracy. Such persons possess authority only because of the positions they occupy in the bureaucratic hierarchy. This is the sole source of their authority and legitimacy. Weber's ideal-typical model of a bureaucracy includes the following characteristics:

1. There is a precise division of labor. The organization owns the job, not the individual incumbent. Each job comes with a clearly defined set of tasks and specifications of the skills required by the job holder.
2. There is a well-defined hierarchy of authority. Job descriptions clearly indicate the authority and responsibilities that go with each job. It is clear who has the authority to ask another to do something, and there are job-related limits on that authority.
3. Job descriptions, including lines of authority, are part of an elaborate system of rules and regulations.
4. In carrying out duties according to the rules and regulations, people act impersonally, according to universal standards.

5. People who hold jobs are organized into specialized units. In particular, the administration, which includes management and ancillary staff like secretaries and bookkeepers, is used to enforce the rules.
6. Jobs represent a career path. People joining the organization expect to move up through the ranks as they gain experience.[3]

Bureaucracy brings with it a distinctive kind of rationality. Rationality of one type or another has existed in all societies, but none have produced the type of rationality distinctive in the United States and Western civilization, what Weber called *formal rationality*. George Ritzer has summarized Weber's use of formal rationality.

> To Weber, *formal rationality* means that the search by people for the optimum means to a given end is shaped by rules, regulations, and larger social structures. Thus, individuals are not left to their own devices in searching for the best means of attaining a given objective. Rather, there exist rules, regulations, and structures that either predetermine or help them discover the optimal methods. Weber identified this as the major development in the history of the world. Previously, people had to discover such mechanisms on their own or with vague and general guidance from larger value systems.[4]

The process of bureaucratization, for Weber, stressed the rationalization of efficiency."[5] As societies became increasingly rationalized, *rational-legal authority*, in which leaders are legitimated by the rule of law, became the dominant organizing principle. This new authority replaced *traditional authority* (authority legitimated by custom) or *charismatic authority* (legitimation of authority achieved when an individual is believed to possess extraordinary, even supernatural, powers).

It was Weber's contention that the modern world was becoming increasingly bureaucratized and human beings were being locked in an "iron cage" of history from which there was no escape. Individuality and creativity were fast disappearing in the face of an unhalting march toward bureaucratization.

Bureaucracy has now become a fact of life in the modern world. During the twentieth century, bureaucratic organization dominates nearly every sphere of human endeavor, a direct outgrowth of the years between the American Civil War and the end of the 1920s, which witnessed marked changes in the structure of America's economy. Economist Gardner Means characterizes this period:

> Mass production and big corporate enterprises took over much of manufacturing; the railroads were consolidated into a few great systems; public

utility empires and the big telephone systems developed; and, even in merchandising, the big corporation played a part.[6]

The federal government has undergone the same process of bureaucratization. In 1870 there were fewer than 50,000 federal employees; today there are almost 3 million. The last three presidential administrations (Reagan, Bush, and Clinton) all began their presidency with a promise to curtail the federal bureaucracy. Reagan and Bush both claimed that government was too big, and both failed to curtail the federal bureaucracy. It is too early to tell whether President Clinton will be more successful, but if recent history is any indication, it does not seem very likely.

The major institutional orders cannot be understood without looking at the role of bureaucracy in society—how it is related to power. Bureaucracies are not neutral but support the interests of those in power who seek to preserve their positions of dominance. Bureaucracy represents the centralization of power, and power is derived from the positions occupied by those at the top. How is power centralized in the economic and political institutions?

The Economic Order

The economic order consists of the ways people organize, produce, distribute, and consume goods and services that they need or want. Goods and services must be both produced and distributed. The accomplishment of these two tasks differs from society to society. In America, goods and services are produced and distributed through the corporation, the dominant mode of economic organization.

Corporations: Economic Giants

A *corporation* is an organization that pursues economic interests. It is a legal entity, separate and apart from its owners. The ownership of a corporation is divided into parts called shares of stock. Persons who own these shares of stock are called stockholders. An important legal feature of the corporation is that it can be dealt with like an individual. It can sue and be sued; buy, sell and exchange property; and engage in a variety of business practices.

In the United States a small number of corporations dominate the economy. For example, 2 percent of the companies in America account for nearly 75 percent of all business. The top 500 industrial corporations, which represent only one-tenth of 1 percent of this 2 percent, control over two-thirds of the business resources, employ two-thirds of the industrial workers, account for 60 percent of sales, and reap 70 percent of the profits.[7]

Since the end of World War II, the largest corporations have grown into multinationals, defined as "a company with its parent headquarters located in one country and subsidiary operations in a number of other countries. The central characteristic of a multinational corporation is that it seeks to maximize the profits not of its individual subsidiaries, but rather of the parent company."[8] Multinational corporations now dominate the world economy, their rise based on two fundamental tenets of modern business ideology: the cult of bigness and the "science" of centralization.[9]

With bigness and wealth comes power, and power has often turned to arrogance in the case of many multinational corporations. They have not hesitated to bribe officials of foreign governments, to intervene in the affairs of sovereign nations, and to have the U.S. government take action favoring them in foreign countries.[10]

Large-scale business in the United States is thus organized around multinational corporate lines. A small number of large corporations own a large share of industrial assets both at home and abroad. Given the vast power of the multinational corporations, it is important to know just who controls them.

Who Controls the Corporations?

For over three decades, from the early 1930s to the early 1960s, the *managerial revolution* thesis was far and away the major view in American society concerning who controlled the corporations. The classic statement of the managerialist position was first put forth by A. A. Berle, Jr., and Gardiner Means in 1932. They contended that the directors or managers of the corporations were in control, that they made the major decisions.[11] Berle and Means argued that until the turn of the twentieth century most large American corporations were controlled by the families of founding entrepreneurs who held the majority of stock in these corporations. However, as these corporations grew larger and larger, it became more and more difficult for the controlling families to maintain their stock control. The control of the large corporations came to rest with "those who [had] the power to select the board of directors."[12]

As stock distribution became more and more widely dispersed, it became apparent that stockholders could not organize themselves into an effective and coherent group that could influence corporate policy. Given this situation, the board tended to become a self-perpetuating oligarchy, free from the influence of the unorganized stockholders. Other managerialists extended this position, arguing that most board members were outsiders from other institutions with part-time appointments to the board and therefore had little knowledge of the day-to-day workings of the corporations. The board members were in no position to contradict policy directives of the managers.[13]

Stock dispersal, according to the managerialists, also helped free the corporation from the restraints of borrowing capital. Capital from the sale of stock and from profits enabled the corporation to repay its debts and forgo further borrowing, which might produce lender influence and restraint. The modern corporation thus was viewed as a large bureaucratic organization dominated by managers who were free not only from the influence of the stockholders but from lending institutions as well.[14]

Beginning in the 1960s, however, critics began to question the managerial revolution position. One of the earliest studies was carried out by Don Villarejo, who showed that as a group, directors of corporations owned sufficient shares to give them financial control over at least two-thirds of the largest 232 corporations.[15] According to Villarejo, a mere 99 of the 2,784 directors of the largest 232 corporations dominated the holdings ($5.2 billion out of $7.1 billion total for the individual directors in their own firms).[16] Other studies, by Burch (1972), Zeitlin (1974), and Kotz (1978), also called into question the managerialist's contention that stock dispersal had restrained managerial autonomy.[17]

Beginning in the 1980s, critics began to assert that the argument over managerialism was misplaced.[18] Even if huge blocks of stocks could be shown to be prevalent in corporate America, possession of controlling interests did not necessarily insure control. The crucial element was the extent to which corporations were dependent upon banks and insurance companies for investment capital. The empirical evidence began to indicate that corporations were indeed tied to the major financial institutions.[19]

This lack of empirical support for managerialism led to the insight by critics that an interorganizational system exists, one that connects firms by a network of corporate ties. The most obvious and common of these ties is via interlocking directorates, or *corporate interlock*, wherein certain individuals sit on a number of boards of directors of the large corporations. Research then took the form of asking the question: Does the small number of elite who direct the major corporations and hold more than one such position of economic power control the corporations?

Corporate Interlock

Mark Mizruchi has defined the research questions on corporate interlock as follows:

First, according to managerialists, as corporations become controlled by insiders, their dependence on, and ties to, other corporations should have declined. Second, if nonfinancials' dependence on financial corporations has declined, then the dominance of financial institutions in the system should also have declined.[20]

However, although stock ownership has become less concentrated, it is not clear that this decline has led to a decline in external control, since the proportion of stock necessary for control is not as high as it once was.

Internal financing increased since the 1960s, and large corporations seemed to be tightly connected in a system of interlocking directorates revolving around financial institutions. Michael Allen, who looked at multiple directorships in large corporations, found that although more managers than inheritors were members of the corporate elite, a large proportion of corporate elites were affiliated with financial institutions and an inordinately large proportion of the elites possessed super-wealth or were members of the upper class.[21] Michael Soref found that individuals whose power came through ownership rather than executive position in the corporation were more likely to have upper-class backgrounds, were more concerned with major decisions, and were able to spread a web of influence to a larger number of corporations.[22]

It is perhaps, this area of financial interlock that has proved to be most fruitful for the critics of managerialism. In particular, the work of Beth Mintz and Michael Schwartz has provided much of the empirical evidence that shows the powerful position financial institutions hold in corporate America.[23] Mintz and Schwartz focused upon two aspects of the interlock process: (1) *structural hegemony* (when the actions of one institution or coordinated group of institutions determine the options available to other institutions and individuals) and (2) whether the interaction in the interlock networks could be identified.[24] Mintz and Schwartz emphasized the importance of *institutional stockholding*. Institutional investing grew out of the first successes of the Congress of Industrial Organizations (CIO) use of pension funding just after World War II to invest as capital.

> Pension monies, coupled with the endowments of foundations and universities and the assets of individual capitalist families given over to professional management, have made institutional investors significant actors in the world of big business. These managers, usually located in the trust departments of the major banks, make investment decisions for their various holdings. They decide what and when to buy and sell.[25]

So widespread is institutional financing that during the 1980s it was estimated that institutional investors accounted for 90 percent of all the stock traded on the New York Stock Exchange and for 67 percent of all stocks held.[26]

Individuals from the major financial institutions sit on the boards of directors of the largest corporations in the United States and influence policy at this level. Mintz and Schwartz documented, how the American business structure is profoundly influenced by a handful of the largest banking institutions of the country. Located mainly in New York City, such giant banks as Chase Manhattan, Morgan and Company, Citicorp, Bankers' Trust New York, and Chemical Bank achieve a loose coordination that usually constrains nonfinancial firms into a conformance that adheres to the interests of the banks themselves.[27] So pronounced was this tendency of corporate interlock that Mintz and Schwartz concluded that:

> Business leadership accrues to a special social type: a cohesive group of multiple directors tied together by shared background, friendly networks, and economic interests, who sit on bank boards as representatives of capital in general. . . . Banks are the primary mechanism for collective decision making within the business sector.[28]

In sum, a small group of individuals is overrepresented in positions of economic power in the United States.

The Political Order

The political order consists of those institutions that are involved with power. Power is traditionally defined as the ability to get others to do one's bidding whether they want to or not.[29] The state or the government is the institution that claims legitimate power and, therefore, is central to understanding power in American society. The two major theories concerning power and the role of the government are examined in order to determine which best describes American society in the 1990s. The two are the Pluralist view and Elitist theory. The latter has two variants: the Marxist view and the power elite view.

The Pluralist View of Power

Pluralism holds that a large number of groups compete over a vast array of interests. Conflict is seen as endemic but is usually smoothed over, because "all active and legitimate groups in the population can make themselves heard at some critical stage in the process of decision."[30] If there are elites, they are quite numerous, highly specialized, and have influence only in their particular sphere of importance. In turn, they are challenged by other elites who are perceived by pluralists as having different interests.

To the pluralist, the government acts as a referee between various interest groups and is always fair in its mediating role. The nation's political leaders, the president, and Congress are seen as being responsive to the general will. The people's demands are registered through elections and are made public through a free press. Though the people do not rule, they select those who do. Government decisions are subject to majority rule and the protection of minority rights. Most decisions thus become compromises that seldom satisfy all parties but are ultimately based on a working consensus where every group has its say and none are dominated. As Seymour Martin Lipset put it, "Democracy in modern society may be viewed as involving the conflict of organized groups competing for support."[31]

The problem with the pluralist view is that it exists only in the pages of high school civics books. Hiding behind the facade of a scientific social science, the pluralists present a biased view of political reality. For example, though they talk about national power structures, for the most part they study only small communities—in particular, one city, New Haven, Connecticut. This investigation has been used to substantiate the view that national power is diffused among competing parties. Political scientist Robert A. Dahl's *Who Governs?* (1961)[32] and two companion volumes by Dahl's former research assistants, Nelson W. Polsby's *Community Power and Political Theory* (1963)[33] and Raymond E. Wolfinger's *The Politics of Progress* (1974),[34] which looked at New Haven in the years 1940–1960, claim that no one ruling elite governs that city. This generalization from one case study is at best poor social science and at worst a prime example of ideological blindness.

Another criticism of pluralism revolves around its naive claim that significant political decisions are made publicly. Political scientist Michael Parenti perceptively pointed out:

> The pluralists overlook the tremendous importance of privately made decisions which might have profound consequences in determining the life chances of millions of people. By so narrowing the scope of inquiry, they conclude that the totality of power is reflected in the visible political arena, in the pursuance only of those decisions which run into enough open conflict as to be defined as "issues."[35]

Given their emphasis upon local levels, pluralists overlook the inability of the social scientist to gain access to important national political decisions. The decisions concerning war and peace, prosperity and slump, are made behind closed doors, and social scientists are not invited.

In short, pluralists present an overidealized picture of the political structure, one not grounded in reality.

Elitist Theory

Social scientists who adhere to an elitist theory of power contend that a small group of individuals make the major political decisions. There are two variants of elitist theory: the Marxist view and the power elite view. The Marxists hold that the super-rich, through their control of major corporations, constitute a tight, cohesive ruling group. Advocates of the power elite theory believe that the ruling group is larger, encompassing not only capitalists but also political and military leaders.

The Marxist View. The Marxist view of power is predicated on the notion that an "upper strata" or "ruling class" runs the United States. Marxist economist Paul Sweezy stated the Marxist case as follows:

> What we have in the United States is a ruling class with its roots deeply sunk in the "apparatus of appropriation" which is the corporate system. To understand this ruling class—its metaphysics, its purpose, and its morals—we need to study, not certain "domains of American life," however defined, but the whole system of monopoly capitalism.[36]

According to most Marxists, the "ruling class" in America uses four mechanisms to control the political process: (1) the special interest process, (2) the process of selecting officeholders, (3) the policy formation process, and (4) the maintenance of ideological hegemony.[37]

1. *The Special Interest Process.* The major corporations are seen as systematically influencing congressional committees and regulatory commissions and agencies. This is accomplished through the workings of lobbyists, back-room superlawyers, trade associations and advisory committees, financing campaigns, granting favors to friendly congressmen, and making sure that members of regulatory commissions have close ties to the very industry they are to regulate. Examples pointed to are how, during the Reagan years, the Environmental Protection Agency (EPA), instead of setting environmental standards for marketing pesticides, let the pesticide industry write its own standards by engaging in a series of informal "decision conferences" between EPA and pesticide representatives. Or how the Occupational Safety and Health Administration (OSHA) adopted standards developed by the oil industry concerning the health safeguards for oil workers rather than its own stringent standards. During the Bush presidency, the Committee of Competitiveness, chaired by then Vice President Quayle and consisting of representatives

of the major industries, thwarted just about every major environmental legislation with the major exception of the watered-down Clean Air Act of 1990.

2. *The Selection of Officeholders.* Candidates for election and high appointive office are seen as being chosen by the upper class. Since the campaign reform acts of the 1970s, corporate and upper-class contributions to the electoral process have been channeled largely through *Political Action Committees* (PACs). The Federal Election Campaign Act of 1974 (FECA) created the basic campaign financial structure that is still in place today. Detailed public reporting of contributions, limits on individual contributions, federal matching funds for presidential candidates, and the creation of a Federal Election Commission (with the power to impose criminal and civil penalties for infractions) were written into law.[38] Limits on how much personal money a candidate could spend was also part of the 1974 Act, but was amended when Senator James Buckley of New York (a member of the upper class and brother of conservative columnist and broadcaster, William F. Buckley) won his case before the Supreme Court, and the Court ruled that it was unconstitutional on First Amendment/Free Speech grounds to limit the total amount of personal money a candidate could spend. This came to a head during the 1992 presidential election when billionaire businessman H. Ross Perot, using his own personal wealth and running as the presidential nominee of a third party (United We Stand), which he created and financed, spent some $65.6 million of his own money and lent another $4.5 million to his campaign. Perot outspent both the Republican nominee George Bush and Democrat nominee Bill Clinton, who were limited by election law to spending $55.2 million.

What has also occurred is that campaign reform, which was intended to limit the impact of business interests, has resulted in just the opposite effect—the strengthening of these interests. Loopholes in the law were found and used to the advantage of the corporate leaders and wealthy individuals. For example, it was legal for independent PACs to pour unlimited amounts of money into the campaigns (for or against) as long as the PAC was not directly or legally connected with a candidate's campaign committee. And a 1979 Amendment allowed PACs to raise unlimited amounts of money for the political parties to help in voter registration and organization-building activities (which could mean almost anything).[39] As William Domhoff succinctly put it: "The effect of the reforms, although unintended, was to corporatize campaign finance and to make political money even more directly tied to the top leaders in the corporate community."[40]

3. *The Formation of Policy.* Most of the major policy decisions made and implemented by the government originate in upper-class con-

trolled policy-formation institutions, such as elite universities, think tanks, and policy-planning groups. Among the most influential policy-planning groups are the Council on Foreign Relations (which formulates basic policies and long term programs for U.S. foreign policy), the Committee for Economic Development, and the Business Council. Marxists see these organizations as being composed of upper-class individuals.

Policy-making groups are dismissed by the Marxists as being predominantly right wing and supportive of the status quo, with no progressive or liberal group having any real impact on policy formation. Although policy is not limited to the interests of the ruling class and sometimes is implemented for the common good (Social Security, Medicare, education reform), on the whole, the policy-formation process is seen as serving the overall interests of the ruling class.

4. *The Maintenance of Ideological Hegemony.* Through its domination of education, religion and mass media, the upper class dominates the popular consciousness. The upper class defines issues for people by seeing to it that only one vision of political-economic reality is presented, a vision that is undergirded by a value system that stresses individualism, the notion that each individual is responsible for his or her lot in life. As Martin Marger states:

> The society's opportunity structure is pictured as open, providing equal chances for all to achieve material success or political power regardless of their social origin. This being so, each individual controls his or her placement in the social hierarchy. Social success, then, is the result of one's willingness to work hard; failure is the result of lack of ambition or desire to improve oneself. Differences in wealth and power are not denied, but they are seen as the product of individual factors rather than the workings of a class system that automatically favors success for the well-born and failure for the poor.[41]

The problem with the Marxist view is that it falls into the trap of conspiracy theory. The upper class is envisioned as being extremely orthodox in its views, with a single-minded determination to see to it that these views are passed down to the people. In reality, there are fundamental splits in the upper class, and it does not always impose a common view on everyone else.

Furthermore, a close look at American society reveals that at least since World War II, the economic order and the political order cannot be understood as one structure. The state is simply not "a committee of the ruling class" as Marxists claim. Power and social inequality are much more complex phenomena than Marxists admit. Given this, the power elite thesis, which plays down the overriding importance of the economic order, would appear to be a more accurate description of power in American society.

The Power Elite View. The power elite thesis, as originally formulated by C. Wright Mills, differs from the Marxist view of power in that Mills carefully avoided the Marxist economic concept of ruling class. Instead, Mills chose a Weberian framework, seeing power as residing in the major institutional orders rather than with the super-rich or the large corporations. To Mills, it was the top leaders in the bureaucratically structured political, economic, and military orders who formed the power elite. The fundamental question asked by Mills was: *What is the nature of the elite in each institutional order?*

First, do the elites in the economic order, or as Mills calls them, the corporate elite, form a distinct social type? Do they perform and incorporate similar role experiences? Looking at the careers of five hundred of the top executives in 1950, Mills found that the executive career was almost entirely a career within the corporate world. Less than 10 percent of the top executives since 1920 entered their positions from independent professional or outside hierarchies. In order to become a member of the corporate elite an executive had to be well liked, an insider who fit in with those already at the top. One not only had to meet the expectations of one's superiors but had to imitate them. Competence was judged by conformity to the values of those already at the top. "To be compatible with the top men is to act like them, to look like them, to think like them; to be of and for them—or at least to display oneself to them in such a way as to create that impression.[42] Therefore, in the economic order, those who wanted to succeed took their cues from their superiors, or the corporate elite, and performed their expected roles within the corporate structure. The end product was a similarity of personality type.

The political order was the same, with political leaders having similar backgrounds. As for the military order, according to Mills, "the harsh initiation at the Point or the Academy—and on the lower levels of the military service, in basic training—reveals the attempt to break up early civilian values and sensibilities in order to more easily implant a character structure."[43]

The personalities of these elites were shaped within the bureaucratic structures they ascended. For Mills, American society was no longer a separate economy, an autonomous political order, and a subservient military order. Historical trends linked these orders, and as they converged there was an overlapping of those individuals who lived out their lives within the milieux that made up the power elite. Their social origins and their common education made them better able to understand and trust one another. Bonds of friendship developed and solidified the power elite. As personal friends, perhaps even neighbors, they met each other both inside and outside the occupational domain. They joined exclusive private clubs, and their children intermarried.

The unity of the power elite consisted of the ease of interchangeability within the bureaucratic structures that characterized the political, economic, and military orders. Donald Regan of Merrill-Lynch became Ronald Reagan's chief of staff; General Brent Scowcroft became George Bush's national security advisor; Richard Rubin of Goldman, Sachs became Bill Clinton's chairman of the National Economic Council. Their expertise in one bureaucratic setting was transferable to another.

In short, Mills saw the power elite as a ruling group in a society characterized by the domination of the political, economic, and military institutional orders. Because they have similar jobs, members of the power elite become similar personality types and ultimately come to see the world in the same terms. It makes little difference whether they are cabinet members, corporation presidents, or generals. Their view of self is shaped by the institutions they come to dominate. The power elite may be in different institutions, but because of the bureaucratization of society they are basically compatible and are easily able to make the transition from one position to another in a bureaucratic structure.

This is not to say that Mill's power elite thesis is without weaknesses. In particular, Mills may have overstated the case for the power of the military and for the cohesiveness of the power elite, and his failure to consider the role of minority groups is an important weakness in his analysis.[44] Of all the criticisms leveled against Mills, perhaps the most telling is that he did not offer enough empirical evidence to support his position. In the years since publication of *The Power Elite*, however, there has been a decided shift in the debate over the locus of power— a shift from the theoretical to the empirical level.

Empirical Research on Power

One of the first social scientists to work in Mills' tradition is G. William Domhoff.[45] Domhoff combines Mills' power elite thesis with a Marxist view of upper-class dominance. In Domhoff's view the upper class rules American society through its holding of key institutional positions.

Using such indicators as the *Social Register*, attendance at elite prep schools, membership in elite social clubs, millionaire status, and legal or corporate success to ascertain upper-class status, Domhoff documented the fact that, overall, the majority of corporation directors are members of the upper class. In order to show how cohesive the members of the upper class are, Domhoff specifically looked at exclusive social and recreational clubs where the corporate and governmental elite met informally. Domhoff claimed that these clubs, the most prominent of which is the Bohemian Club outside of San Francisco, help to foster cohesiveness among the power elite as they provide settings for informal discussions of national policy. According to Domhoff:

> Constant interaction in small-group settings leads to the social cohesion that is considered to be an important dimension of a social class. This social cohesion does not in and of itself demonstrate that members of the upper class are able to agree among themselves on general issues of economic and governmental policy. But it is important to stress that social cohesion is one of the factors that makes it possible for policy coordination to develop.[46]

Domhoff argues that upper-class individuals also hold key positions in the political order, enabling the business world to assert a large amount of control over the government. He examined key cabinet posts from 1932 to 1964 and found that five of eight secretaries of state and treasury and eight of thirteen secretaries of defense were members of the upper class.[47] Domhoff assumes that the military, given its emphasis upon graduation from the service academies, was not upper-class dominated and did not delve into the social characteristics of the military leaders.

This preponderance of upper-class people in important governmental positions was further reinforced by other studies. Peter Freitag looked at the relationship of the executive branch to big business by focusing upon the 205 individuals who held U.S. Cabinet positions from 1897 to 1973. He found that at least 76.1 percent and possibly as many as 87.7 percent of the Cabinet secretaries had corporate affiliations.[48] Beth Mintz, in a parallel study of the background characteristics of all Cabinet members during the same time period, found that almost "90 percent of all cabinet members who held office . . . were members of either the social or business elite."[49]

Another important study of the relationship of the economic order to the political order was carried out by Michael Useem in the early 1980s.[50] Useem begins with the premise that both the family and managerial capitalism of the past have given way to a new form of classwide capitalism, what he calls *institutional capitalism*. As an important aspect of institutional capitalism, Useem postulates the existence of an *inner circle* of senior managers of large corporations whose multiple board memberships (interlocking directorates) enables them to view the world in terms of the long-term interests of business as a whole. This inner circle is primarily located in the major financial institutions, and variations of it are found to exist both in the United States and Great Britain. Useem's inner circle members are more unified and more politically active than other business leaders and are thus able to transcend the narrow interests of their own corporations or industries.

The inner circle members, consisting of a limited number of senior managers of the nation's largest firms who are also involved in the affairs of the large corporations, are propitiously situated for the role they come to play in economic and social policy. They are more unified than other business leaders, "for the inner circle shares a special culture, informal acquaintanceship, and common tradition more developed than anywhere else in the business community."[51] This enables the inner circle to possess a unique social cohesiveness.

> Social cohesion implies that the inner circle is truly a circle; acquaintanceship networks are dense, mutual trust and obligation are widespread, and a common sense of identity and culture prevail. All these features are embodied and reinforced in a variety of social institutions, ranging from clubs to intermarriage. While social cohesion is not a necessary precondition for mobilization, it is a powerful facilitator.[52]

The inner circle also shares distinctly different views from its fellow directors and managers. This difference occurs because members of the inner circle have a better understanding of the overall opinion of the large corporations on matters concerning contemporary economic policy and of the complexities and intricacies of the political environment in which business operates.[53] As a result, the inner circle members have a better sense of how the political process works and are thus better able to maneuver in it and affect it than are those with more narrow self-interests.

The inner circle thus comes to constitute a leadership cadre for the entire business community. It shapes corporate political activity but it does so in combination with the political environment of business. The state may well change or add to the social policies desired by the inner circle, but according to Useem, the inner circle constitutes a powerful force for shaping the policy of the United States. It may not always get what it wants from the government, but the wishes and desires of the inner circle of business leaders must always be taken into account whenever social policy is undertaken.

Perhaps the most comprehensive study of the locus of power in American society is that of Thomas Dye in his important work, *Who's Running America?* (1986).[54] Dye, working squarely within the tradition of C. Wright Mills, begins with the assumption that power lies in the large institutions. He then develops an operational definition of what he refers to as *institutional elites*, "those individuals who occupy *the top positions in the institutional structure of American society.*"[55] Dye's view of the institutional elite is an exceptionally broad one. For him the elite is located not just in the political, economic, and military orders, as Mills claimed, or in the economic and political orders, as Domhoff and Useem hold, but in twelve sectors: (1) industrial corporations, (2) utilities, transportation, and communications, (3) banking, (4) insurance, (5) investments, (6) mass media, (7) law, (8) education, (9) foundations, (10) civic and cultural organizations, (11) government, and (12) the military.

Dye's findings concerning the power of an institutional elite are mixed, with much of his research supporting the elitist argument but with enough data that does not fit to lend some credence to the pluralist argument. In support of the elitist position, Dye found that only 5,778 individuals occupied the top 7,314 positions he labeled as the institutional elite. This indicated that there were multiple holdings of top positions by individuals, in some cases six, seven, and eight positions. All in all, 15 percent held more than one position and are what Dye called "interlockers." Furthermore, some 32 percent of all top *positions* were interlocked with other top positions. Dye labeled individuals in these positions as an *inner group* of the nation's institutional leaders.[56]

Second, one hundred of the top five hundred industrial corporations were family owned or controlled. Since the top five hundred corporations accounted for over half of all industrial assets, half of all banking assets, and two-thirds of all insurance assets, the role of the upper class, while not as all-powerful as the Marxists would have us believe, nonetheless was substantial. Indeed, Dye found that approximately 30 percent of the institutional elite could be considered upper class. When we take into account that this class represents about 1 percent of the population, we can see how disproportional is its representation in the power structure. Third, Dye was led to the conclusion that the banking and finance institutions hold strategic ownership positions in the institutional elite, thereby supporting the contention of Mintz and Schwartz and Useem.

Fourth, like C. Wright Mills and the other elite theorists, Dye holds that the elite are socially cohesive. Members of the elite interact with each other and know each other socially. "They come together not only in multiple corporate boardrooms, but also at cultural and civic events, charitable endeavors, foundation meetings, and university trustee and alumni get-togethers. They are also members of the same exclusive social clubs—the Links, Century, Knickerbocker, Burning Tree, Metropolitan, Pacific Union."[57]

Although Dye's findings tend to support the elitists' argument more than the pluralists', nevertheless, he did find that interlocking of directors has declined moderately since the early 1970s. In 1970, he had estimated that about 20 percent of all top leaders were "interlockers." In 1980, this estimate was only 15 percent.[58] This means that 85 percent of the institutional elite were affiliated with only one position, which does not provide great support for the interlocking directorate assumption. In fact, Dye found very little overlap among people at the top of the corporate, governmental, and military sectors of society. To the extent that high government officials were interlocked at all, it occurred only in the civic, cultural, and educational institutions. And finally, although Dye found evidence of consensus rather than competition as characterizing elite opinion, he did find fractionalism among the elite.

Dye sums up his research as follows:

> Our findings do not fit neatly into either the elitist or the pluralist leadership model. The fact that roughly 7,000 persons in 6,000 positions exercise formal authority over institutions that control over half the nation's resources is itself an indication of a greater concentration of power. But despite institutional concentration of authority, there is considerable specialization among these 7,000 leaders.[59]

Although researchers working within the elite tradition, with the notable exception of Dye, had fairly well established the existence of interlocking directorates and their location in the major financial institutions (something which even Dye agreed with), pluralists have rightly called for a demonstration that these interlocking directorates produce common political action. What was needed was an empirical demonstration that showed just exactly what were the behavior consequences of interlocks. However, by 1979, only one study had demonstrated any link between interlocks and corporate political behavior. An unpublished doctoral dissertation by Thomas Koenig found that interlocked corporations were more likely than less-interlocked ones to have contributed to Richard Nixon's 1972 reelection campaign.[60]

With this dearth of empirical research in mind, elitist theorist Mark Mizruchi, in *The Structure of Corporate Political Action* (1992), set out to document the behavioral aspects of corporate action.[61] Mizruchi studied the factors that generated political unity and conflict among a sample of fifty-seven of the largest manufacturing firms in the United States in an attempt "to identify the conditions under which political unity and opposition occur within business."[62]

Mizruchi started from what most researchers believe: that firms' political behavior corresponds when they have similar interests. He went beyond this, however, to show that a network of economic and social ties among firms, what he calls, "mediating mechanisms," provided an important source of similar political behavior. These mediating mechanisms, according to Mizruchi, provide a better explanation for business unity in the political realm than do common interests alone.

Focusing specifically on congressional testimony and political contributions by the top managers of the firms in his sample, Mizruchi
concluded that direct interlocks between firms were a significant predictor of agreement in congressional testimony among these firms but
are only a marginally significant predictor of contributions to the same
political candidates. The number of ties that firms shared with the same
financial institutions, however, was the best predictor of similarity in
political behavior. But even firms with indirect ties with financial institutions were more likely to engage in similar political behavior than were
those firms tied to one another.[63]

Mizruchi's findings lend support to Useem's claim that an inner
core of business leaders exists and exerts a large amount of political
pressure. What Mizruchi added is that it is "not size or concentration
per se, but rather corporations' economic, organization, and social interaction that unify and empower them."[64]

In sum, the recent evidence concerning power in American society,
though not completely conclusive given the absence of any strong empirical evidence that those in positions of power are overwhelmingly upper-
class members, as the Marxists hold, still does tend to support elitist
theory rather than pluralism. C. Wright Mills' view that power is to be
found in the major institutional orders (with the exception of the military
order) seems to present a more accurate portrayal of political reality than
the pluralist version of a balance of power among competing interest
groups, or the Marxist staple of the state being a "committee of the
ruling class." The economy is centralized and controlled by multinational
corporations; the political system is dominated by a small group of individuals who have ties to the economic order. Those who are in the top
positions in the political and economic institutional orders, what are
referred to here as the dominant institutional orders, hold power in American society.

In chapter 7, the subordinate institutional orders—religion, education, the family, and the mass media—are examined to show how they
support and legitimize the two dominant orders.

Summary

This chapter has shown that the economic and political orders dominate
American society. The United States is seen as a bureaucratized, centralized political economy. Power resides with the government and the multinational corporate structure.

The economic order was seen as being dominated by the multinational corporations, which are large-scale bureaucratic entities. It was
also shown that interlocking directorates exist within the corporate realm

and, for the most part, produce a fairly cohesive group of business leaders. Power within the corporate structure was seen as being centralized in the financial institutions.

The political order, in particular pluralism and elitist theory and its two variations, Marxist theory and the power elite thesis, were discussed and analyzed. A modification of C. Wright Mills' power elite theory (power resides with a small group of influential individuals who occupy the major positions in the political, economic, and military institutional orders) to exclude the military order (seeing it as an appendage of the state) was seen as the most accurate explanation of political power in the United States. The economic and political orders are the dominant orders in American society—the other orders merely serve to legitimate them.

Notes

1. C. Wright Mills, *The Power Elite* (New York: Oxford University Press, 1956), p. 6.

2. Ibid., p. 7.

3. Max Weber, *The Theory of Social and Economic Organization*, Talcott Parsons, ed. (New York: Free Press, 1964), pp. 329–41.

4. George Ritzer, *The McDonaldization of Society* Newbury Park, CA: Pine Forge Press, 1993, p. 19.

5. Weber, *The Theory of Social and Economic Organization*.

6. Gardiner Means, "Economic Concentration," in Maurice Zeitlin, ed., *American Society, Inc.* (Chicago, IL: Markham, 1970), p. 5.

7. Michael Schwartz, "Introduction," in Michael Schwartz, ed., *The Structure of Power in America* (New York: Holmes and Meier, 1987), p. 3.

8. Ronald Muller, "The Multinational Corporation and the Underdevelopment of the Third World," in Charles K. Wilber, ed., *The Political Economy of Development and Underdevelopment* (New York: Random House, 1973), p. 125.

9. Richard Barnet and Ronald E. Muller, *Global Reach: The Power of the Multinational Corporation* (New York: Simon and Schuster, 1974), p. 37.

10. See Michael J. Parenti, *Democracy for the Few*, 6th ed. (New York: St. Martin's Press, 1995).

11. A. A. Berle, Jr., and Gardiner Means, *The Modern Corporation and Private Property* (New York: Macmillan, 1933).

12. Ibid., p. 66.

13. See Alfred Chandler, *The Visible Hand: The Managerial Revolution in American Business* (Cambridge, MA: Harvard University Press, 1977); and John K. Galbraith, *The New Industrial State* (New York: Signet, 1968).

14. Mark S. Mizruchi, "Managerialism: Another Reassessment," in Michael Schwartz, ed., *The Structure of Power in American Society* (New York: Holmes and Meier, 1987), pp. 7–33.

15. Don Villarejo, "Stock Ownership and Control of Corporations," *New University Thought* 2 (Fall 1961 and Winter 1962):33–77, 47–65.

16. Ibid.

17. Philip H. Burch, Jr., *The Managerial Revolution Reassessed* (Lexington, MA: D. C. Heath, 1972); Maurice A. Zeitlin, "Corporate Ownership and Control: The Large Corporation and the Capitalist Class," *American Journal of Sociology* 79(1974):1073–1119; and David Kotz, *Bank Control of Large Corporations in the United States* (Berkeley: University of California Press, 1978).

18. See, in particular, works by Mark S. Mizruchi, *The American Corporate Network, 1904–1974* (Beverly Hills, CA: Sage, 1983); "Managerialism: Another Reassessment"; and *The Structure of Corporate Political Action* (Cambridge, MA: Harvard University Press, 1992).

19. Linda B. Sterns, "Corporate Dependency and the Structure of the Capitalist Market," Ph.D. diss. State University of New York at Stony Brook, 1982.

20. Mizruchi, "Managerialism: Another Reassessment," p. 12.

21. Michael Patrick Allen, "Continuity and Change within the Core Corporate Elite," *Sociological Quarterly* 19(1978):510–21.

22. Michael Soref, "Social Class and a Division of Labor within the Corporate Elite: A Note on Class, Interlocking, and Executive Committee Membership of Directors of U.S. Firms," *Sociological Quarterly* 17(1976):360–68.

23. Beth Mintz and Michael Schwartz, *The Power Structure of American Business* (Chicago, IL: University of Chicago Press, 1985); and "Sources of Corporate Unity," in Michael Schwartz, ed., *The Structure of Power in America* (New York: Holmes and Meier, 1987), pp. 16–33; and "Corporate Interlocks, Financial Hegemony, and Intercorporate Coordination," in Schwartz, ed., *Structure of Power in America*, pp. 34–47.

24. Mintz and Schwartz, *Power Structure of American Business*.

25. Mintz and Schwartz, "Sources of Intercorporate Unity," p. 19.

26. Ibid.

27. Mintz and Schwartz, "Corporate Interlocks, Financial Hegemony, and Intercorporate Coordination," p. 35.

28. Mintz and Schwartz, *Power Structure of American Business*, p. 254.

29. Weber, *Theory of Social and Economic Organization*, p. 152.

30. Robert A. Dahl, *A Preface to Democratic Theory* (Chicago, IL: University of Chicago Press, 1956), p. 137.

31. Seymour Martin Lipset, *Political Man* (Garden City, NY: Doubleday, 1960), p. 34.

32. Robert A. Dahl, *Who Governs? Democracy and Power in an American City* (New Haven, CT: Yale University Press, 1961).

33. Nelson W. Polsby, *Community Power and Political Theory* (New Haven, CT: Yale University Press, 1963).

34. Raymond E. Wolfinger, *The Politics of Progress* (Englewood Cliffs, NJ: Prentice-Hall, 1974).

35. Michael J. Parenti, *Power and the Powerless* (New York: St. Martin's Press, 1978), pp. 29–30.

36. Paul Sweezy, "Power Elite or Ruling Class," in G. William Domhoff and Hoyt B. Ballard, eds., *C. Wright Mills and the Power Elite* (Boston, MA: Beacon Press, 1968), p. 129.

37. The following description of these processes is based on Albert J. Szymanski and Ted George Goertzel, *Sociology: Class, Consciousness and Contradictions* (New York: Van Nostrand, 1979), pp. 204–9; and Arnold K. Sherman and Aliza Kolker, *The Social Bases of Politics* (Belmont, CA: Wadsworth, 1987), pp. 170–75.

38. W. Lance Bennett, *The Governing Crisis: Media, Money, and Marketing in American Elections* (New York: St. Martin's Press, 1992).

39. Ibid.

40. G. William Domhoff, *Who Rules America?* (Englewood Cliffs, NJ: Prentice-Hall, 1983), p. 123.

41. Martin N. Marger, *Elites and Masses: An Introduction to Political Sociology*, 2nd ed. (New York: Van Nostrand Reinhold, 1987, pp. 310–11.

42. Mills, *The Power Elite*, p. 141.

43. Ibid., p. 193.

44. For a discussion of the major criticisms of the power elite thesis, see Joseph A. Scimecca, *The Sociological Theory of C. Wright Mills* (Port Washington, NY: Kennikat Press, 1977), pp. 88–97.

45. See, for example, G. William Domhoff, *The Power Elite and the State: How Policy Is Made in America* (New York: Aldine De Gruyter, 1990); *Who Rules America Now?* (Englewood Cliffs, NJ: Prentice-Hall, 1983); *The Powers That Be: State and Ruling Class in Corporate America* (New York: Random House, 1979); *The Bohemian Grove and Other Retreats: A Study in Ruling Class Cohesiveness* (New York: Harper and Row, 1975); and *The Higher Circles* (New York: Random House, 1970).

46. Domhoff, *Who Rules America Now?* p. 50.

47. Ibid.

48. Peter Freitag, "The Cabinet and Big Business: A Study of Interlocks," *Social Problems* 23(1975):137–52.

49. Beth Mintz, "The President's Cabinet, 1897–1972: A Contribution to the Power Structure Debate," *Insurgent Sociologist* 5(1975):135.

50. Michael Useem, *The Inner Circle: Large Corporations and the Rise of Business Political Activity in the U.S. and U.K.* (New York: Oxford University Press, 1983); and "Business and Politics in the United States and the United Kingdom," *Theory and Society* 12(1983):281–308.

51. Useem, *Inner Circle*, pp. 9–12.

52. "Business and Politics," p. 290.

53. Ibid.

54. Thomas R. Dye, *Who's Running America?* (Englewood Cliffs, NJ: Prentice-Hall, 1986).

55. Ibid., p. 10.

56. Ibid., p. 61.

57. Ibid., p. 168.

58. Ibid., p. 169.

59. Ibid., p. 184.

60. Thomas Koenig, "Interlocking Directorates among the Largest American Corporations and Their Significance for Corporate Political Activity," Ph.D. diss., University of California, Santa Barbara, 1979.

61. Mizruchi, *Structure of Corporate Political Action*.

62. Ibid., p. 32.

63. Ibid., p. 243.

64. Ibid., p. 254.

CHAPTER 7

The Subordinate Institutional Orders: The Family, Religion, Education, and the Mass Media

The role of the subordinate institutional orders—the family, religion, education, and the mass media—is to legitimate the domination of the political and economic orders and to socialize individuals to accept the status quo. Political scientists call this process "political socialization." The perspective offered here is a variant of the traditional political science view, which focuses on such political attributes as voting, acceptance of political authority, belief in patriotism, and so forth. This chapter will be concerned with how and why most individuals come to identify with the status quo by what has been called a "hegemonic theory" of domination or control. *Hegemony* is a term used "to describe the domination that selected interests in a society exercise over the whole of society."[1]

> In capitalist countries the ownership of private property is a major source of political domination. Groups that own the industrial base of society (factories, transportation systems, communication networks, raw materials, etc.) are able to dominate nonowning groups. Moreover, it is in the interest of the dominant groups to get the nondominant groups to accept domination. In the capitalist example, it is in the interests of property owners to convince nonproperty owners that private ownership of basic production and services in society is both important and legitimate. If nonowners accept that private ownership is right and appropriate they are in effect accepting the rules and values by which they are dominated.[2]

Hegemonic theory starts with the assumption that the unequal distribution of valued goods and services, or social inequality, is biased in favor of dominant or ruling groups. In other words, there are always the same winners and the same losers, and the losers must accept that the way things are (why they lost) is natural or appropriate or legitimate. This occurs through socialization.

Hegemonic theory is similar to what Karl Marx called "false consciousness," a condition in which large numbers of individuals accept a social and political order when it is not in their objective interest to do so. An important question to ask is, "What are the factors which constitute the basis of legitimacy and political integration?"[3] What is generally considered to be the best answer to this question has been provided by Max Weber. According to Weber, the social order is upheld through *authority*, the belief that the exercise of power is legitimate.[4]

Authority

Weber accepted the premise of inequality and then asked the question, Why doesn't conflict over this inequality break out more often than it does? His answer was that power was institutionalized, that there is an important distinction between legitimate and illegitimate power. Legitimate power is based on the belief that certain rules and commands are binding and, more important, that it is *desirable* to obey such rules and commands. Those in power must motivate others to accept them as leaders. If those in positions of authority are thought to be legitimate, they can issue commands and take other actions that are followed voluntarily and without question. If the people of a society are socialized to accept the norms attached to authority structures and can be induced to obey them voluntarily, then commands issued by legitimate political leaders will be followed unhesitatingly. The result is an ordered society.

In such an ordered society the mass of people come to believe that their interests are being served by their leaders and that the distribution of power and economic goods is just and fair. Legitimacy masks inequality and the conflict that would arise if there was not this acceptance.[5]

When power is hidden it can be exercised more efficiently. And since legitimacy is the most effective device for hiding power, the single most important aspect of socialization becomes the acceptance of the power structure as legitimate.[6] A simple principle holds here: "The greater success dominant groups have in nurturing and reinforcing belief in the legitimacy of a system, the less resistance they will face in the exercise of their power."[7]

As we saw in chapter 5, the individualistic perspective is fundamental to legitimizing social inequality. The four subordinate or minor institutional orders will now be examined to help us understand the part they play in socializing people to accept the individualistic perspective, thereby perpetuating a hegemonic theory of domination.

The Family

The family is of paramount importance in any society. The overwhelming majority of people spend more time, effort, energy, and emotion in family life than in any other social institution.

There are three kinds of family organization: the *nuclear* family, the *one-parent* family, and the *extended* family. Although the nuclear family is being challenged in sheer numbers by the one-parent family, it is still both the predominant and the ideal family structure in American society. It consists of a husband, a wife, and their dependent children. The core of the family is based on marriage, and because of this, the nuclear family is sometimes also referred to as the *conjugal* family. (*Conjugal* means "to unite in marriage.") Most people are members of two different types of nuclear families during their lifetimes. The nuclear family unit into which we are born is called the *family of orientation*. The family we marry or enter into is called the *family of procreation*.

Due to the prevalence of teen-age, unmarried pregnancies and the rise in the "three D's"—*death, divorce* and *desertion*—the one-parent family has become the fastest growing type of family structure in the United States. The one-parent family traditionally can be found most often in the lower classes and is female headed. More recently, it is increasing in the middle class, and men are beginning (albeit a small beginning) to head one-parent households.

The third type is the extended family. It consists of several generations of blood relatives living together. The extended family is also called the *consanguine* (meaning having blood ties or the same ancestors) family. This kind of family is no longer common in the United States. When it is found, it predominantly occurs among such ethnic groups as Italians, Greeks, Mexicans, or Asians.

The family is the prime agency of socialization and as such is expected to reproduce the cultural patterns of society in the individual. It is therefore naive to think that the family is a counteracting agent working against the demands of the society. Children learn norms and social rules that are defined by external agents.

> If the reproduction of culture were simply a matter of formal instruction and discipline, it could be left to the schools. But it also requires that culture be embedded in personality. Socialization makes the individual want to do what he has to do; the family is the agency to which society entrusts this complex and delicate task.[8]

The power elite uses the family to insure social control. One way it accomplishes this is by making consumerism an important function of the family. Consumerism is tied to individualism by equating buying with success. Through consumerism, through purchasing, individuals are taught that they can achieve success that escapes them in other areas. The family, taking its cue from the dominant institutional orders, socializes its members in this direction.

Consumerism and the Family

The industrial revolution had a profound effect on the American family, transforming it from a producing unit to a consuming unit. Before industrialization, most Americans lived on farms or in small towns. Every member of the family had a job to perform. The father worked the farm or the small shop. The mother took care of the household duties, cooked and preserved all the food, and made all the clothing. Each child shared in the work, whether it was helping the father plow the fields or helping the mother with food canning or other domestic chores.

With industrialization came a changeover from working on the farm to working in the factory. Where before, families were units in which each person worked and helped one another, family members were now hired as individuals; labor was exchanged for wages. Authority outside of the family had to be accepted, and individual accountability for work came to be stressed. The factory took over from the family as the basis of social organization. "Stripped of internal necessity the family was weakened, left to the cohesion of emotional bonds."[9] The internal authority of the family eroded more and more as individuals came to buy rather than produce.

Advertising would help spread even further the emphasis upon consumption, entwining it with the ideology of individual ability. One could achieve success, according to advertisers, by consuming products equated with the successful. One could dress a certain way and drive a particular car, and then one could have personal and economic success. By diverting attention away from the class structure, the ideology of individual achievement supported the status quo.[10]

Women and Consumerism

In the early days of industrialization, men, women, and children left the home to work in the factory. Reformist movements, however, quickly saw to it that child labor was abolished or at least curtailed. Children were to stay in the home and attend school for part of the day. Consequently, a pattern developed in which men went to work and women remained in the home to care for children. Women's work was housekeeping, sewing, laundry, food preparation, and other chores that could be done in the home.[11]

Housework was time consuming and dull. To offset this state of affairs, consumerism promised glamour through the purchase of beauty products. This promise could be transferred to the daughter, providing a belief in a better future, if not for oneself, then for one's offspring. As one ad put it:

> "From her first smile" use Palmolive Soap. "Correct skin care starts in infancy. It is the duty that every mother owes her child." The rewards for adopting such an educational philosophy were in tune with the demands of the age. "Schoolgirl complexions come now as a natural result. . . . To assure your child's having one through the years, you must take 'proper' steps now."[12]

The transcendence of one's mundane existence could be accomplished by purchasing. Buying thus becomes the means of giving value to oneself.

Achieving the Good Life

A widely accepted belief today is that the more one purchases, the more successful one is. Work in a modern bureaucratized society like the United States has become more and more specialized and more and more meaningless. The primary function of work has become a means to an end. It provides the means to purchase, if not success, at least the illusion of success. If one cannot be a member of the upper class or the middle class, at least one can consume similar items. Credit allows more purchasing power, and the illusion of social mobility is ensured; the open society is shored up by consumerism. The family as a primary agency of socialization incorporates consumerism into the education of children.

As the family was transformed from a productive unit to a consuming unit, it saw its authority wither. Authority now is located outside the family; youths no longer look primarily to their parents for guidance but often see them only as sources of the money that enables them to buy the products they see advertised. The consumer-oriented society treats youths as a lucrative market to be manipulated into following the latest fads. It is no coincidence that the growth of the purchasing power of youths coincided with the entrance of women into the labor market. Advertisers use the feelings of guilt of parents who do not spend as much time as they once did with their children (due to work demands) to get them to compensate through purchases.

The family is in the business of socialization, but because socialization must support big business, families cannot be trusted to carry on in their own way. The family, as a subordinate institution order, legitimizes the status quo, one that is consumer oriented.

Religion

Religion has been defined differently by numerous sociologists. For the purpose of a common ground for discussion, Ronald Johnstone's definition of religion as "a system of beliefs and practices by which a group of people interprets and responds to what they feel is supernatural and sacred"[13] will be used.

Hans Gerth and C. Wright Mills have said that the main features of any religion can be grasped if the following facts are known:

> Its attitude toward the status quo, its representative class or leadership; the sources of its religious authority; its type of religious assembly and organization; the chief end of life it holds out; its views of the superhuman and of life after death; its sexual code; its magical features, if any; and its attitudes toward politics, work and education.[14]

In light of this definition, the three major religions of the United States—Protestantism, Roman Catholicism, and Judaism—will be examined.

Protestantism

Protestantism consists of numerous denominations and sects, and to provide answers to the questions posed by Gerth and Mills, requires a consensus that may not be there. However, there are certain similarities, and we shall concentrate on the most apparent.

Protestantism generally supports the status quo. Its leaders, particularly Episcopalian and Presbyterian clergy, who are usually upper or upper-middle class, are given high social status in American society. The source of religious authority is the Bible, which is interpreted by individual members. The religious organization and assembly of Protestantism is almost as diverse as the denominations and smaller sects of which it is comprised. The purpose of life is individual salvation and the promise of an afterlife for the faithful. In sexual matters, the liberal denominations—such as Episcopalian, Presbyterian, Lutheran—are more open, while the fundamentalist churches—such as Baptists, Assembly of God, or Seventh Day Adventists—are strict. Nearly all Protestant denominations, with the possible exception of the Jehovah Witnesses and similar groups, place great stress on mass education of the laity. Protestantism has been a bulwark of support for the political structure, and its puritan strain has emphasized the importance of hard work and success.

Roman Catholicism

The Roman Catholic church, like its Protestant counterparts, strongly supports the status quo. Religious authority is derived from the

Bible, which is interpreted by the clergy under the sanction of the Pope. The Pope is seen as being in a direct line of succession from St. Peter, who was told by Christ to found His Church. The assembly and organization of the church are built on a strict hierarchy beginning with the Pope as head of the Church and ending with priests at the bottom. Nuns are excluded from this hierarchy, and the Catholic church has thus far refused to admit women to the priesthood. Laymen participate in the Sunday mass, but most have little other contact with the church's hierarchy. Roman Catholics believe in heaven, hell, and purgatory (most Protestants believe only in heaven and hell), and the chief goal is individual salvation. The sexual code is quite strict, condemning premarital and extramarital sex, contraception (other than natural methods), and abortion. Roman Catholics believe in miracles and venerate saints as heroes and heroines. The church's attitude toward politics is conservative. Education is stressed for clergy and laymen, and the work ethic is as strong as it is among Protestants.

Judaism

Given the history of persecution among Jews, Judaism is less supportive of the political order than are Protestantism and Catholicism. Its representative leaders are Torah teachers and rabbis, and religious assembly and organization are based in the temple. Judaism's chief end of life rests upon faith in being chosen by God for a covenant fellowship. There is no idea of personal resurrection after death, only the return of the soul to its source. Among the orthodox, its sexual code is as strict as any puritan denomination, but the more liberal beliefs resemble more closely those of some of the major Protestant denominations. Education is highly valued, perhaps more so than in any other religion. The work ethic also is ingrained in the Jewish tradition; the Protestant Ethic in American society is definitely not limited to Protestants.

As can be seen from these capsule descriptions, organized religion is basically supportive of the political status quo. This relationship needs to be investigated.

Religion and the Power Structure

Religion can be related to power and the dominant institutional orders in at least three ways.[15] First, to the extent that religious norms coincide with political norms, religion performs an integrating function by reducing tensions that result from inequality. Religion can justify unequal success and may even encourage submission to the state. The relationship between religion and the power structure is basically one of support.

A second possible relationship occurs when dominant institutions assume control over religion and use it as an instrument of coercion. Religion then becomes a tool of the ruling elites. Marx's view of religion as the "opium of the masses" provides an adequate description of this possible relationship. The historical role that the Roman Catholic church played in supporting social inequality in Latin America is an example of this type of relationship between religion and the dominant orders.

There is a third possibility. A sharp tension may develop between religion and the state. The norms and requirements of religion may contradict and challenge those of the ruling elite. Primary allegiance to an overriding religious system can transcend political boundaries and undermine political demands. Revolution can occur if the religious ferment is strong enough. The Iranian revolution is an example of this possibility.

Of the three possible relationships, the first resembles most closely conditions in the United States. How does this process work in American society? How does religion legitimate the dominant institutional orders?

Religion, Legitimation, and Capitalist Society

One of the most important ways religion legitimates American society is by its reinforcement of the capitalist ethic. Again, we turn to the insights of Max Weber. In his classic work, *The Protestant Ethic and the Spirit of Capitalism* (1904),[16] Weber proposed the thesis that without Protestantism (particularly Calvinism), capitalism would not have appeared in Europe.

Capitalism requires hard work and a rational approach to the accumulation of money. But the making of money must not be an end in itself; capital must be reinvested to earn more capital. This approach to economics, Weber argued, stemmed from the "Protestant Ethic" of hard work and deferred gratification. The early Calvinists believed in predestination—that God had already decided whether a person would achieve salvation in Heaven or be condemned to Hell. Only a small minority were among the "elect" who would go to Heaven, and nothing could be done to change the fate of the damned. The sacred duty of the believer was to abstain from pleasure and instead to work for the glory of God. In order to alleviate their anxiety, people needed a sign that they were among the chosen, and this sign was "worldly success." The more successful individuals were at their work the more likely they were to be among the elect. And since profits should not be spent on pleasure, they were to be reinvested. In this way, modern capitalism was born.

Other religions did not provide the same fertile ground for capitalism's growth. In the Roman Catholic tradition, for example, when Adam and Eve ate of the forbidden tree of knowledge, their punishment was banishment from the Garden of Eden. Henceforth, they and their de-

scendants would have to earn their food by the "sweat of their brows"—
by working. Work was thus initially a punishment in the Catholic tradition. Hence, the reason for the flourishing of capitalism in Protestant
rather than Roman Catholic countries.

Nor were the Eastern religions conducive to capitalism's development. Hinduism, with its belief in reincarnation to a lower status for
those who try to leave their present caste, provided no reinforcement
for a belief in upward mobility, so necessary for capitalism. Buddhism
stresses mysticism rather than earthly goals, and Confucianism emphasizes a static social structure. All these religions thus discouraged capitalism.

Although there have been criticisms of Weber's thesis,[17] as well
as a debate among conservative Christian thinkers over the support of
capitalism,[18] the consensus is that the religious institution in the United
States strongly supports the capitalist economic system.

Religion and the Radical Right

Perhaps the most important way in which religion supports the
dominant orders is through the connection of fundamentalist religious
denominations with radical right-wing political beliefs. A prime example
was the 1992 Republican presidential campaign in which the religious
right was one of the most powerful factions in the campaign and essentially wrote the Republican party platform. Ultraconservative religious
leaders rallied fundamentalists, who were concerned about liberal trends
in contemporary society and what they saw as a threat to their values,
to the more conservative Republican party. The individualist perspective, or emphasis on individual salvation, manifested itself in a corres-
ponding disapproval of social action and welfare programs, thereby
matching the radical right's laissez-faire economic and political
ideology.[19]

The Protestant fundamentalists have been successful in translating
this disapproval into policy by concentrating upon local elections. In
particular, school boards have become a prime target as fundamentalists
have rallied against what they consider to be the evil of secular humanism
and have sought to expunge this evil from the schools. In a recent book,
Joan Delfattore describes the world view of Protestant fundamentalists
as follows:

> In their way of looking at life, all decisions should be based solely on the
> Word of God; using reason or imagination to solve problems is an act of
> rebellion. Everyone should live in traditional nuclear families structured
> on stereotyped gender roles. Wives should obey their husbands and chil-
> dren their parents, without argument or question.

Regarding the world outside the family, the United States has, since its inception, been the greatest nation on earth. Any criticism of its founders, policies, or history offends God and promotes a Communist invasion by discouraging boys from growing up to fight for their country. Since war is God's way of vindicating the righteous and punishing the wicked, anti-war material—and by extension, criticism of hunting or gun ownership—is unpatriotic, disrespectful to God, and detrimental to the moral fiber of American youth.

Pollution and other environmental concerns are humanist propaganda designed to provide an excuse for international cooperation, either of which is capable of destroying this country. Unregulated free enterprise represents God's will and the American way of life. International cooperation might lead to a one-world government, which would be the reign of the Antichrist and bring about the end of the world. Besides, world unity would have to be based on religious tolerance, which is unthinkable.[20]

Along with the issue of abortion, the control of local school boards and curricula promises to be a major battleground for the Christian right in the Clinton administration.

Education

Public schools are organized to preserve the status quo. As with the family and the religious order, the socialization process is a legitimizing process. By stressing an individualistic perspective, schools socialize children to accept the reasoning that failure or success is solely dependent upon their own ability. Given that the self is a social creation, the acceptance of the individualistic perspective becomes essential for the acceptance of the legitimacy of the schools. The schools are a primary agent of socialization, perhaps second only to the family, and have a profound impact on how individuals develop a self-image. Obviously, those persons who have developed a positive self-image based on their success in school would be extremely reluctant to question the ideology of individual ability. To do so would mean the questioning of an integral part of their image of self. For example, teachers, the socializing agents with whom students have the most day-to-day contact in the schools, have their jobs because they were academically successful and accepted the socialization process. When they get up in front of a class, they almost instinctively impose these internalized values upon their students. Teachers and students alike are playing roles, and their behavior centers on the expected norms.

It is of paramount importance that lower-class individuals in particular come to see their failure as caused by their own lack of ability. If they thought otherwise, they would not believe that the system is a legitimate one and that equal opportunity has been provided for them. Individuals who accept the role of educational failure based on their own shortcomings do not look to the society and its class structure for the source of their problems. Instead, they look inward. They are not aware that the overwhelming majority of those who are successful in the educational institution are so because their parents were highly educated.

To be fair, teachers and principals do not purposely denigrate their students, locking them into a role of failure. They, too, are products of a system enmeshed in a network of power and status relationships. The basic problem of public education is that it is conservative; that is, it conserves or supports whatever power arrangements control the society.

Lower-class students are socialized in the schools to accept their lower-class status. They are discriminated against, and the schools do little to help them succeed. When they do succeed, they must overcome a number of hardships. The successful lower- or working-class student usually must be academically the best of the lower class, whereas the average middle- and upper-class student is "guaranteed" success.

Because lower-class students and their parents are powerless and suffer from a poor self-image, which they internalize as their own fault, they eventually come to see themselves as less than worthwhile human beings. They become second-class citizens, devoid of autonomy and control over their own lives. As one student of education put it:

> We have managed to engage the brighter students, to alienate them (sometimes inadvertently) from their original group, and to entice them into the kingdom by the promise of acquisitive success. But the vast majority of ghetto dwellers remain uncalled and unreaching; powerless and ultimately hopeless. The large society decided to whom and to how many the right of choice will be offered. This is quite different from the notion (supposedly "basic" to American ideology) that choices and freedom are the rights of all.[21]

Millions of lower-class children are characterized by powerlessness and a depressed self-image. Unable to articulate their own interests and concerns, the lower classes are "socialized into compliance. . . . They accept the definitions of political reality as offered by dominant groups, classes and governmental institutions."[22] Alternatives to the system are precluded and *distorted communication*, a process in which "all forms of restricted and prejudiced communication that by their nature inhibit a full discussion of problems, issues, and ideas that have public relevance,"[23] becomes the rule.

The schools work to insure the legitimacy of the society. Lower- and working-class youths, because they are taught to accept their failure as caused by their own individual inabilities, blame themselves for their state rather than the school system and the society it represents. They accept the system because "distorted communication" does not permit them to construct any defense against the legitimation of the dominant orders.[24] It is extremely difficult to mount an argument against the dominant institutional arrangements when one has had very little experience in doing so. The result is very little political reflection as the socioeconomic arrangement of the society basically is accepted. The schools are in the business of political socialization (hegemonic domination), and they do an excellent job in supporting the status quo.

Political Socialization

A simple definition of political socialization has been offered by Richard Dawson, Kenneth Prewitt, and Karen Dawson, and it will be used here: "Political socialization is concerned with the personal and social origins of political outlooks."[25] Schools have generally taken upon themselves the task of socializing future citizens. The following statement by a superintendent of schools when speaking about immigration illustrates this acceptance:

> The public school is the greatest and most effective of all Americanization agencies. This is the only place where all children in a community or district, regardless of nationality, religion, politics, or social status, meet and work together in a cooperative and harmonious spirit. . . . The children work and play together, they catch the school spirit, they live the democratic life. American heroes become their own, American history wins their loyalty, the Stars and Stripes, always before their eyes in the school room, receives their daily salute. Not only are these immigrant children Americanized through the public schools, but they, in turn, Americanize their parents, carrying into the home many lessons of democracy learned at school.[26]

In reality, though, what goes on in the schools is somewhat different, for political socialization is more closely related to domination and control than to learning the lessons of democracy.

The doctrine of individual ability is centered in the power arrangement of the society. Subordinate group members must believe that they have an opportunity to achieve a dominant position. The socialization process is a political one, and the values of individual success and failure are directly linked to the political values taught in the schools. To question the one is to question the other, and the schools seek to contain such questioning. In short, the schools are organized around the process of persuading subordinate groups to accept the social values that insure their own subordination.

Parents are not likely to provide political education for their children, other than an attachment to a particular political party. Politics are not given a high priority in the typical American family. Parents make little effort to articulate their own positions, and they do not push their children toward adopting them. Exposure to political issues occurs usually in current events discussion in the elementary school or in a high school civics course. This absence of political education by the family has made it easy for schools to step into the breach.

Schooling, Patriotism, and Authority

The school curriculum is potentially one of the most important primary instruments of political socialization. All students receive formal instruction in the nature and glory of the established order. James Coleman[27] offered a useful distinction between *civic training* and *political indoctrination*. Civic education is that part of political education that emphasizes how a good citizen participates in the political life of the nation. Political indoctrination, on the other hand, concerns learning a specific political ideology intended to rationalize and justify a particular regime. Although the official position of the school purports to emphasize the former to the exclusion of the latter, in practice the distinction between the two is often blurred. In particular, this can be seen by focusing upon the political activities that the student comes into daily contact with in the classroom.

From the first day in kindergarten the child is exposed to such rituals as saluting the flag, singing patriotic songs, and honoring national heroes. The importance of these ceremonies becomes obvious when the amount of time and resources allocated to them is considered. Teachers are compelled, either by law or strong social norms, to spend countless school hours on classroom activities that stress patriotic values.

These rites are important politically for a number of reasons. First, rituals can be seen as the acting out of a sense of awe toward what is symbolized by the ritual. Basic feelings of patriotism and loyalty are instilled by the expression of devotion. "The feelings of respect for the pledge and the national anthem are reinforced daily and are seldom questioned by the child."[28] Submission, respect, and devotion are associated with these acts. The rituals "establish an emotional orientation toward country and flag though an understanding of the meaning of the words and actions has not been developed. These seem to be indoctrinating acts that cue and reinforce feelings of loyalty and patriotism."[29]

Second, schools maintain those few political loyalties that have been established in the family and introduce new conservative ones. Third, and perhaps most important of all, rituals emphasize the collective

nature of patriotism. Saluting the flag and singing the national anthem
are group activities.[30] "We-feelings" of a specific nature are reinforced.
There is a general feeling of identification with others who are part of
a benevolent whole. Nationalism, partisanship, and identification with
a legitimate political group have more meaning when they are experienced as part of a collectivity and the ritual life of the school revolves
around such a collective experience.[31]

The teacher is the major school socializing agent for the student.
The teacher represents the authoritative spokesperson for society, for
the teacher is usually the first model of political authority encountered
by the student. The special quality of the teacher's authority can be seen
by a comparison between the parent and the teacher.

When children respond to their parents as authority figures they
do not separate the role and the person. Mother is mother, and no one
else can be mother. Parents are thus very personal authority figures,
dispensing both rewards and punishments in what can easily appear to
be idiosyncratic and perhaps even capricious ways. The teacher, on the
other hand, is much more of a political authority figure. Children learn
that they must obey whomever happens to be in the role of teacher,
regardless of who it is. The teacher, like the police officer, mayor, or
president, is a part of a larger institutional order that demands the
child's obedience.

In short, the ideals expressed about political socialization—that it
educates citizens to take an active role in a democratic society; that
children who come from backgrounds where their parents are apathetic
or unknowledgeable about political matters are afforded the opportunity
to acquire knowledge about the real workings of the political system,
which then enables them to engage in serious thinking about politics
and social problems—do not exist in practice. Instead, the picture that
emerges from studies of political socialization supports the contention
that the primary purpose of schools is to legitimate the established social order.

Textbooks and Conformity

Textbooks are bland affairs, reflecting a picture of a smoothly working political system with little or no conflict and social inequality. Controversy is something that is to be guarded against. A case in point concerns
how elementary and secondary textbooks are adopted in the state of
Texas, where a state-wide Board of Education chooses the texts that are
to be used throughout the state.

The state-wide board, because of the size of the market, has told
publishers that they will not receive approval unless state education offi-

cials are allowed to rewrite whole paragraphs in proposed textbooks.[32] Given that by Texas educational law, capitalism must be extolled and socialism and communism must be portrayed negatively, there is no critical analysis of the economic system permitted. Educational Research Analysts, a tax-exempt, ultraconservative Texas based organization, founded by Mel and Norma Gabler, for the past two decades has provided detailed reviews of all textbooks under consideration by Texas schools. And the Texas Board of Education has accepted the Gablers' ideas about educational materials as being what Texas parents want, giving the couple a tremendous impact upon what is published or not published.

For example, the Gablers' strong opposition to socialism has led them to attack a favorable presentation of the New Deal. Giving in to pressure from Educational Research Analysts, the board made a publisher change the sentence "These [New Deal] programs and policies were generally successful in restoring the prosperity of many Americans" to "In spite of this, by the time the United States entered World War II, most of the nation was still suffering from the Great Depression."[33]

What is at work here is an interlocking of public education with the conservatism of religious fundamentalism. When asked why they got into the textbook criticizing business, Mrs. Gabler said, "I believe that this is what God would want us to do."[34] Although the Gablers stated that they were not censors, that they merely provided information, they and others like them have a chilling effect on textbook publishing. Rather than offend anyone's sensibilities, textbook publishers, in pursuit of profits, steer clear of controversy. Consensus becomes the byword. A certain pattern clearly emerges from the literature on political socialization; schools teach legitimacy.

The Mass Media

The average American is exposed to the mass media (television, newspapers, newsmagazines, and radio) to a far greater extent than they are exposed to any other institutional order, with the possible exception of the family. Because of this, the mass media is in a pivotal position to shape the consciousness of the American people and to insure, along with the other subordinate institutional orders, that legitimacy is accepted. Let us take a look at the three principal components of the mass media—advertising, television, and news dissemination—in order to see how the mass media accomplishes this.

Advertising

To be fully understood, advertising must be seen as a direct response to the needs and values of the power structure. The purposes of advertising are to persuade a person to consume, thereby supporting the corporate elite, and to control the person, thereby supporting the societal arrangement. Concerning this latter purpose, if people are unhappy with the society and their place in it, advertising attempts to put this unhappiness to work for stability. The road to success, as was noted previously, lays in changing oneself. Stuart Ewen, in his book on the history of advertising, goes even further, claiming that "the control of the masses required that people, like the world they inhabited, assume the character of machinery—predictable and without any aspirations toward self-determination."[35]

People are objects to be manipulated. They are a market, no more, no less. Social passivity is stressed. Advertising holds up consumption as an antidote to boredom and social frustration. Perfume, for example, can turn a tired housewife into someone ready for a night out on the town. A certain brand of clothing can make a man irresistible to women. Equality, too, is purveyed through consumerism. Advertisements seek to blur class lines by showing that if people cannot achieve upper-class status they can at least consume the same products as those above them in the class hierarchy.

Individual ability or failure is always stressed in advertising. Social failure is reduced to individual failure as individuals are told to look to themselves first. Is it bad breath? Use a particular mouthwash and success is yours. Or maybe it's dandruff. A dandruff shampoo will help you make the right first impression with the opposite sex.

Solutions to major social problems are also seen in individual terms. Air pollution is a problem that can be solved by an eyewash; family problems can be solved by the use of a certain brand of pain killer. We are constantly being manipulated, our behavior controlled by advertising and the mass media it is a part of. Buy, buy, buy, is the message conveyed. Salvation lies in consumption. Commodities define our social existence.

Television Programming and Social Control

Television spews forth a basic ideology of control and legitimation. Television shows "encourage viewers to experience themselves as anti-political, privately accumulating individuals."[36] With few exceptions programs convey images of social steadiness. In show after show, plots express a world immune to structural social change. It is individuals who must somehow change themselves if they are to overcome what little adversity is presented. Regularly scheduled shows do not involve heroes or heroines who challenge the core values of the power structure.

Even when the characters are portrayed as struggling to make ends meet, as not being middle class or upper class, as in the *Roseanne* show, it is their individual initiative that helps them to solve their problems. When Dan Conners' motorcycle repair shop goes under (a casualty of the recession of the late 1980s and early 1990s) and Roseanne loses her job, Roseanne's mother comes through with a gift of $10,000, which is used to start a new business. Inculcated with the entrepreneurial spirit, Roseanne will succeed. Everything will always come out all right at the end, and the implication is that so too should the viewers' problems. They too can find a solution that does not disturb the status quo.

When blacks are depicted in prime-time television, the reality that so many live in poverty and one-parent families is rarely if ever shown. The Bill Cosby Show, which went off the air in 1991, was one of the most popular shows in the history of American television. It portrayed a family headed by a doctor and a lawyer, whose children's problems could be solved by a comedic version of "father knows best." So, too, with another popular show, *The Fresh Prince of Bel Air,* which revolves around the antics of a poor black kid from Philadelphia who is living with his ultra rich aunt and uncle in upper-class Bel Air, California. The Huxtables and the Fresh Prince can make it, and so can every black. Structural inequality plays no role— it is individual initiative that defines success or failure.

The soap operas, in their own way, also provide a message in support of the status quo. Given the mostly lower- and working-class viewing audience, and the upper- and middle-class professionals who are the major characters of the daytime screens, one could easily argue that soap operas represent a version of the old cliches "the rich are unhappy" and "the rich have worse problems than the poor." Be content with what you have—things could be worse, you could become too successful.

Sports on television provide still another example of social control. Listeners are bombarded with statistics and facts; listeners are treated symbolically as people who know what is going on in the world. Sports become a world one can readily grasp. Indeed, in modern American society, the use of trivial statistics is often mistaken for knowledge and deep understanding.

Sociologist and former athlete Harry Edwards offered the following summary of the role of sports in American society.

> The circus of sport offers not only social stability but balm for individual stresses and anxieties. The sport fan, for example, finds that the success of his favorite team or athlete reinforces his faith in those values that define established and legitimate means of achievement. He returns from the game to his job or his community reassured that his efforts will eventually be successful. So when he is cheering for his team, he is actually cheering for himself. When he shouts "kill the umpire!" he is calling for the destruction of all those impersonal forces that have so often hindered his own achievement.[37]

Whether the message is presented during prime time, in daily soap operas, or in sports broadcasts, it is the same—social stability and control.

The News

The American people receive news of the world around them through television news broadcasts, daily newspapers, weekly news magazines, and radio newscasts. Radio is no longer as important as it once was for the dissemination of news.

The accepted wisdom is that reporters and newscasters function under the "freedom of the press" and are "guardians of truth" and objectively report the facts of the news without regard to political or social pressures. Lee Bollinger, a law school dean described the ideal image of freedom of the press as follows:

> In the United States the government is forbidden by virtue of the First Amendment from censoring or punishing the press for what it chooses to say. The press is not licensed. . . . It need not clear with the government what it proposes to publish. And, except under very limited circumstances, the government may not punish the press for what it has said. Libel, invasions of privacy, extraordinary threats to national security, and a few other justifications may permit government to limit press freedom, but these circumstances are limited and scrutinized closely by the courts. It is, indeed, the function of the courts to protect press freedom against government interference and to decide when those rare instances arise.[38]

A close look at the news media will show that the accepted image is highly exaggerated.

Who Owns the Press?

The journalist A. J. Liebling once sarcastically said that the freedom of the press is for those who own the presses. In the United States, the mass media is owned by large corporations.

> Eight corporations control the three major television networks (NBC, CBS, ABC), some 40 subsidiary stations, over 200 cable TV systems, over 60 radio stations, 59 magazines including *Time* and *Newsweek*, chains of newspapers including the *New York Times*, *Wall Street Journal*, *Los Angeles Times*, and *Washington Post*, 41 book publishers, and various other media enterprises.[39]

Although reporters and newscasters consider themselves to be objective in their pursuit of stories, the owners and editors are under no such restraint. Rupert Murdock, the owner of one of the world's largest media conglomerates, was once asked about the extent of his influence on the editorial policy of his newspapers, given his well-known political conservatism. Murdock responded: "The buck stops on my desk. My editors have input, but I make final decisions."[40] This is not to say that the free press is a myth, that it is totally controlled by the owners, but only to point out that the decision-making process is often controlled by individuals who have their own agenda. Reporters produce investigative reports that question the status quo, but they must struggle not only to get the information for their stories but also to see to it that their stories see the light of day. No less a figure than Walter Cronkite, who until his retirement in 1980 was considered the most influential newscaster in the United States, states that "my lips have been buttoned for almost twenty years. . . . CBS news doesn't really believe in commentary."[41]

Bias in News Reporting

Although reporters often try to be as objective as possible, their own personal biases enter into their stories, producing a subtle form of social control. One of the more obvious biases that permeates the dissemination of the news is the manner in which news reporters deal with the class system in America. Income differences among people are rarely mentioned. Class, class groupings, and class differences are rarely considered newsworthy. Exceptions concern foreign countries, which are seen as having clearly demarcated class structures, unlike the United States. According to Herbert Gans, when the news media does mention America's stratification system:

> It recognizes four strata: the poor (now sometimes called the underclass), the lower-middle class, the middle class, and the rich. "Lower-middle class" is the journalistic equivalent for what sociologists generally call the working class; "middle class" usually refers to the affluent professionals and managers sociologists call upper-middle class. . . . The class system in the news leaves out the sociologists' lower-middle class: the skilled and semi-skilled white-collar workers who are, next to blue-collar workers, the largest class in America, however they are labeled. In eschewing the term "working class," the news also brings blue-collar workers into the middle class; and by designating upper-middle-class people as middle class, it makes it appear to be more numerous than they actually are.
>
> Journalists shun the term "working class" because for them it has Marxist connotations, but even non-Marxist notions of class conflict are outside the journalistic repertoire of concepts.[42]

Public officials are rarely if ever distinguished by class background. For the most part, reporters tend to attribute middle and upper-middle status to almost everyone.

Although journalists like to consider themselves as objective reporters of what they see, most carry with them certain beliefs that play a part in their reporting of the news. Foremost among these values, and one which as we have seen is essential for hegemonic domination, is the belief in individual ability.

Reporters usually seek out stories of individuals struggling against and overcoming more powerful forces. Self-made men and women are always attractive to the news media. That there are so few of them among the over 250 million people in this country is overlooked. By failing to focus on the forces and circumstances that mitigate against achieving success for the overwhelming majority of lower- and working-class people, the news media perpetuate the myth that opportunity is always available for those who are willing to try hard enough. Aren't the handful of self-made men and women they report on convincing evidence of this, they covertly ask?

For the most part, national journalists, who are solidly upper-middle class themselves, have slanted the news to insure the preservation of an upper-class and upper-middle-class social order. As pointed out in the discussion on education, to do otherwise would ask them to question their own image of self-worth.

For Every Watergate

Reporters thus are not quite the guardians of objective truth they purport to be. For every Woodward and Bernstein, who determinedly pursued the truth behind Watergate, there were countless thousands of reporters and editors who were eager to accept the line that the event that led to the resignation in disgrace of President Richard Nixon was merely "a second-rate burglary" carried out by underlings.

There are numerous incidents where those in control of the media refused or were afraid to report the truth when it went against the power structure. In the 1940s, Theodore H. White, who was later to become famous for his series of books on the making of the president, reported accurately that the Chiang Kai-shek government in China was corrupt and would not survive. Henry Luce, owner and publisher of *Time* and *Life*, then the most powerful news organs in the country, and a friend of Premier Chiang, refused to publish White's stories in *Time*. White was later fired. He was considered "too left wing" by other editors.[43] CBS was no better. When William Shirer, a leading political commentator of the 1950s, became critical of the Truman loyalty program for governmental employees and of the Truman doctrine for Greece and Turkey, he was shifted out of prime time.[44]

The news media were also less than courageous during the Joseph McCarthy era. Few journalists reported McCarthy in context or held him to what he had said the day or week before. Editors and journalists, for the most part, hid their love of truth when McCarthy made noise.

It was not until Vietnam that the news media began to function effectively in the role of objective reporters of information. And, although initially reporters took a critical stance, the record of their editors was less than proud. According to Pulitzer Prize winning journalist David Halberstam:

> The best reporters of the war . . . had their stories neutralized by the wire services, whose dispatches, based on the [government] briefings, were printed with surprising regularity on the front pages of the nation's great newspapers. Some reporters complained bitterly to their editors, telling them to bury the briefing story inside. To little avail.[45]

Unfortunately, the gains made during the coverage of the Vietnam war may have been eroded. In particular, coverage of the Persian Gulf war was so controlled that Hedrick Smith, a former Moscow bureau chief for the *New York Times* concluded that in reporting the war

> reporters protested that they were being forced to work under similar or worse conditions than those my compatriots and I had known in Moscow. . . . The main view of the conflict in Iraq . . . consisted of the videotapes the Pentagon chose to release. All facts were sanitized to suit the Defense Department and then funneled through a narrow tube of official briefers.[46]

The Pentagon's rationale for the strict control of news and images from the Persian Gulf was made in the name of national security and to save lives. But it is doubtful that telling such truths as 70 percent of all weapons parts used by American troops in the war were foreign made or that the bombings and killing of thousands of Iraqi soldiers as they desperately tried to retreat might have been inhumane (as came out later)[47] would have contributed to heavier American casualties. Instead, the facts could only have affected the public image that the Pentagon wanted the American people to see. In short, freedom of the press became a casualty to image. Power often defines what passes for truth.

Summary

This chapter looked at the manner in which the four subordinate institutional orders—family, religion, education, and mass media—legitimate the status quo and how they are part of a "hegemonic theory" of domination or control, a situation in which selected interests hold sway over the whole society.

The family is the primary agency of socialization. Socialization entails the acceptance of "consumerism" and "individualism." Consumerism encourages members of the family to buy in order to establish their self-worth. Individualism becomes enmeshed with consumerism via advertising. To be successful one need only use a particular product. In this manner, attention is diverted away from the class structure.

Religion also supports the legitimacy of the society. After a brief look at Protestantism, Roman Catholicism, and Judaism, the role of religion in politics was examined. Religion was shown to support the capitalist system through a strong belief in the "Protestant Ethic." A relatively new phenomenon in which religion supports the status quo is the relationship between Christian fundamentalist religious belief and radical right-wing political theory. The individualistic (individual salvation) emphasis in Christian fundamentalism was shown to manifest itself in a disapproval of social action and welfare programs, which matched the radical rights' laissez-faire economic and political ideology.

Education is an important part of the legitimation process. Public schools stress that the failure of the lower classes is solely dependent upon their lack of ability. The ideology of individual ability diverts attention away from the power structure. Schools teach students to look inward for the locus of their problems. With this looking inward, students are taught to accept the prevailing political and social order. Political socialization guarantees that alternatives to the status quo are rarely if ever discussed.

Lastly, the mass media was analyzed. It was shown that advertising, television programming, and the news media serve a legitimizing purpose in the society. Advertising turns people into objects to be manipulated and controlled. Television programming also is geared toward social control, as an image of social order and stability is presented on the screen. The news media, which are thought to be the guardians of truth, were seen as somewhat less than that. The upper-middle-class biases of owners, editors, journalists, and newscasters produce a tendency to support the power structure. Again, individualism was seen as an important part of reporters' and broadcasters' belief systems. By focusing upon the few individuals who do achieve, the impression is given of an open society, something which is not the case.

Notes

1. Richard E. Dawson, Kenneth Prewitt, and Karen S. Dawson, *Political Socialization*, 2d. ed. (Boston, MA: Little, Brown, 1977), p. 24.

2. Ibid., p. 25.

3. Clauss Mueller, *The Politics of Communication* (New York: Oxford University Press, 1973), p. 5.

4. The following section is based on Max Weber, *The Theory of Social and Economic Organization*, Talcott Parsons, ed. (New York: Free Press, 1947), pp. 324–86. For a clear and concise secondary analysis of Weber's use of the concept "authority," see Robert Bierstedt, *Power and Progress* (New York: McGraw-Hill, 1974), pp. 242–59.

5. James T. Duke, *Conflict and Power in Social Life* (Provo, UT: Brigham Young University Press, 1976), p. 68.

6. Ibid., p. 245.

7. Mueller, *Politics of Communication*, p. 131.

8. Christopher Lasch, *Haven in a Heartless World* (New York: Basic Books, 1977), p. 4.

9. Stuart Ewen, *Captains of Consciousness* (New York: McGraw-Hill, 1976), p. 117.

10. For an historical analysis of this phenomena see ibid., pp. 113–23.

11. Albert J. Szymanski and Ted George Goertzel, *Sociology: Class, Consciousness, and Contradictions* (New York: Van Nostrand, 1979), p. 288.

12. An ad in the *Ladies Home Journal* of Oct. 1925 cited in Ewen, *Captains of Consciousness*, p. 176.

13. Ronald L. Johnstone, *Religion and Society in Interaction* (Englewood Cliffs, NJ: Prentice-Hall, 1975), p. 20.

14. Hans Gerth and C. Wright Mills, *Character and Social Structure* (New York: Harbinger Books, 1964), p. 242.

15. J. Milton Yinger, *The Scientific Study of Religion* (New York: Macmillan, 1970).

16. Max Weber, *The Protestant Ethic and the Spirit of Capitalism*. Talcott Parsons, trans. (New York: Scribners, 1958).

17. See Richard Tawney, *Religion and the Rise of Capitalism* (New York: Harcourt, Brace, 1976); J. Milton Yinger, *Religion and the Struggle for Power* (Durham, NC: Duke University Press, 1946); and Richard Quinney, *Providence* (New York: Longman, 1980).

18. Craig M. Gay, *With Liberty and Justice for Whom? The Recent Evangelical Debate over Capitalism* (Grand Rapids, MI: Eerdmans, 1991).

19. Walter H. Capps, *The New Religious Right: Piety, Patriotism and Politics* (Columbia: University of South Carolina Press, 1990); Ted G. Jelen, *The Political Mobilization of Religious Belief* (Westport, CT: Praeger, 1991); and Matthew C. Moen, *The Transformation of the Christian Right* (Tuscaloosa: University of Alabama Press, 1992).

20. Joan DelFattore, *What Johnny Shouldn't Read: Textbook Censorship in America* (New Haven, CT: Yale University Press, 1992), pp. 36–37.

21. Stanley Charnofsky, *Educating the Powerless* (Belmont, CA: Wadsworth, 1971), p. 7.

22. Mueller, *Politics of Communication*, p. 9.

23. Ibid., p. 119.

24. Clauss Mueller, "On Distorted Communication," in Hans Peter Dreitzel, ed., *Recent Sociology*, no. 2 (New York: Macmillan, 1972), p. 107.

25. Dawson, et al., *Political Socialization*, p. 1.

26. Quoted in Robert A. Dahl, *Who Governs?* (New Haven, CT: Yale University Press, 1961), pp. 317–18.

27. James S. Coleman, ed., *Education and Political Development* Princeton, NJ: Princeton University Press, 1965), p. 226.

28. Robert P. Hess and Judith V. Torney, *The Development of Political Attitudes in Children* (Garden City, NY: Doubleday/Anchor, 1968), pp. 123–24.

29. Ibid., p. 124.

30. Dawson, et al., *Political Socialization*, p. 148.

31. Ibid., pp. 148–49.

32. DelFattore, *What Johnny Shouldn't Read*, p. 138.

33. Ibid., p. 149.

34. "Was Robin Just a Hood?" *Time*, Dec. 31, 1979, p. 76.

35. Ewen, *Captains of Consciousness*, p. 84.

36. Todd Gittlin, "Prime Time Ideology: The Hegemonic Process in Television Entertainment," *Social Problems* 26(1979):253.

37. Harry Edwards, "The Black Athletes: Twentieth Century Gladiators for White America," *Psychology Today* 7 (Nov. 1973), p. 4.

38. Lee C. Bollinger, *Images of a Free Press* (Chicago, IL: University of Chicago Press, 1991), p. 1.

39. Michael Parenti, *Inventing Reality*, 2d ed. (New York: St. Martin's Press, 1993), p. 26.

40. Ibid., p. 33.

41. Quoted in ibid., p. 5.

42. Herbert J. Gans, *Deciding What's News* (New York: Pantheon, 1979), p. 24.

43. David Halberstam, *The Powers That Be* (New York: Knopf, 1979), p. 85.

44. Ibid., p. 133.

45. Ibid., p. 453.

46. Hedrik Smith, ed., *The Media and the Gulf War* (Washington, DC: Seven Locks Press, 1992), p. xiii.

47. See Stuart Auerbach, "U.S. Relied on Foreign-Made Parts for Weapons," and Steven Coll and William Branigin, "U.S. Scrambled to Shape View of Highway of Death," in Smith, *The Media and the Gulf War*.

CHAPTER 8

Intervention and Change

Contemporary society mitigates against individuals becoming free and autonomous human beings. People are objects to be manipulated, and social institutions strip individuals of mastery over their own lives. Roles are defined for us by others, and these roles often discourage any flexibility of self.

In their quest for scientific respectability, positivist social scientists apply natural science research methods to the social world. Mainstream sociology thus becomes another in a long series of ideologies that treat human beings as objects devoid of freedom, as entities dependent upon external forces.

To counteract these conditions individuals need a means through which they can come to know how and why they are being manipulated and controlled. This is the promise of humanist sociology. It assumes that individuals *want* to be free and *can* be free in a society that tries to socialize them to not be free.

The starting point of a humanist sociology is an investigation of how people are limited by social institutions, how powerful individuals who occupy positions in these structures make decisions in their own best interests and to the detriment of most everyone else. Humanist sociology seeks to have an impact upon people's lives and therefore searches out points of effective intervention in the structures that shape human lives. In this manner humanist sociology holds out the option of control over one's destiny and one's social milieu.

Toward a Humanist Sociology

A humanist sociology must be able to explain the interrelationship between self and society, to synthesize a dynamic view of self with a workable notion of social structure. By combining the thoughts of Ernest Becker and C. Wright Mills, two of the greatest humanist social scientists of this century, such an integrated humanist sociology is available. Because chapters 5, 6, and 7 concentrated on Mills' notion of social structure, Becker's view of the self, especially the importance of self-esteem, is reexamined in this chapter.

Becker and the Concept of Self-Esteem

Becker assumed that human beings are self-reflexive, that is, they are able to think about their own past actions. Self-reflexivity enables individuals to think about their actions, to *act* rather than *react*. Individuals are able to control their actions; it is possible to possess a strong self-image or have high self-esteem. However, according to Becker, few people truly realize this state. Because of the socialization process in American society, far too many children learn that they are valuable and loved only if they do not act but merely respond.[1]

Parents constrict their children (most often inadvertently) by not allowing them to test the potential of their own powers against the reality of the external world. Children fail to learn how to tolerate frustrations, how to separate themselves from objects, how to achieve an independence from their environment. The result is constricted children and, subsequently, constricted adults. Children become overly dependent on external things for support.[2] To gain self-esteem, children shape themselves into people who rely on continued parental support and approval, tailoring their actions to suit their parents' wishes. Becker stated:

> It is an inner self-righteousness that arms the individual against anxiety. We must understand it, then, as a natural systematic continuation of the early efforts to handle anxiety; it is the durational extension of an effective anxiety-buffer. We can then see that the seemingly trite words "self-esteem" are at the very core of human adaptation: they do not represent an extra self-indulgence, or a mere vanity, but a matter of life and death. The qualitative feeling of self-value is the basic predicate for human action.[3]

Self-esteem, for Becker, was the key to understanding people; it is the dominant motive for human behavior. Without it individuals literally break down. Life is a constant struggle to establish mastery over one's life, a mastery that can bring an adequate measure of self-esteem. The problem with American society is that its pervasive materialism hinders individuals from achieving the self-worth they need and desire. An example is the rampant consumerism discussed in chapter 7; Americans are taught that their sense of self comes from what they buy. Objects are more important than people. Children are routinely punished for dirtying tablecloths, breaking glasses, scratching furniture. The message received is that people are less valuable than things.

Self-esteem must be transferred from the external world and put firmly under one's own control.[4] In order to take this step toward autonomy, the individual needs self-knowledge. Indeed, it is a truism to state that human beings are *inauthentic* when they do not know why they do what they do.[5]

It is a sad, but true, commentary that most people live out their lives in institutions in which they have no voice and perform tasks that have no meaning for them. They become automatons, "cheerful robots," to use C. Wright Mills' descriptive term[6]—people who dare not question the tasks they routinely perform. Individuals derive no sense of control over their lives and have no feelings of autonomy. Indeed, Becker defines mental illness as just this state of being—a constriction of self-powers. "The typical psychiatric syndromes are simply failures to carry through self-satisfying action in the face of new, problematic situations."[7]

Instead of promoting an individualism that leads to freedom, American society promotes an individualism that produces control. Those who make up the power structure do not want free and autonomous citizens, for such people would not tolerate the leaders who now govern. A citizenry composed of free and autonomous citizens, Becker sarcastically wrote,

> would be something to reckon with because they would place most of those who hold prominent political office today back in the real-estate offices, restaurants, cigar and clothing stores which they never should have left, were it not for the fantasies of the masses. Think of how natural it would be to see the faces that we are now obliged to watch for hours on TV back behind the counters where they belong.[8]

The obvious question is, How can a society that promotes self-esteem be ushered in given what we know about the structure of American society?

Bringing About Social Change

American society is dominated by a small group of people who occupy positions of power. As we saw in chapter 6, C. Wright Mills refers to these people as a "power elite." What is important is not what they are called but that a small number of individuals exercise control over a society that does not allow the vast majority of individuals to develop to their full capabilities.

The first step toward changing this arrangement involves an awareness of the conditions that constrict people. The chapters on social organization, social inequality, and dominant and subordinate institutions demonstrated this purpose and showed that oppression is institutionalized in American society. We live in a society that is not as open as we are told, or as poverty free, or as just as the popular image holds. To change society, the question should be asked: *Why do people accept this state of affairs?*

The principal reason that people accept their position, their lot in life, is because of the acceptance of the legitimacy of various institutions. However, withholding legitimacy would make the society and the power structure that controls it problematic. If enough people reject the legitimacy of those above them, there is little that power holders can do outside of using force. The use of force means that subordinate individuals carry out the implementation of force. And if there is a rejection by subordinates of the authority and legitimacy of those issuing orders to use force, the powerful cannot carry out their own orders.

The withholding of legitimacy is, of course, much easier said than done. Those individuals who tell the boss what he or she can do with their jobs quickly find themselves attempting to collect unemployment insurance. What is needed are organizations or groups of individuals who are aware of society's constrictions and who seek autonomy for themselves and others. For example, if one student in a class feels that he or she is being oppressed by the professor, there is little the student can do under the contemporary educational hierarchy. However, should all or most of the students in a class feel this way, they become a group that must be reckoned with. Professors can fail one, or two, or even a few students, but they could not survive for long if they failed entire classes for several semesters. Having been stripped of their legitimacy by their students they would soon lose their authority.

Authority is vested primarily in the dominant institutional orders: the political and the economic. Legitimacy is vested in the subordinate orders: the family, religion, education, and the mass media. If people are to eventually live in a humane society, then humanist sociology must locate those points within the present society that are most conducive to change. Many might not agree with the next statement; however, I feel strongly that *the political and economic institutions are too strong and too entrenched in the social hierarchy to be changed.* Therefore, if a viable restructuring of society is to take place it must come through the withholding of legitimacy by the family, education, religion, and the mass media.

In the preceding chapter, I showed how the subordinate institutional orders are used as instruments of social control. However, expressed throughout this text is the premise that we are not only creatures in but also creators of the social reality within which we live out our lives. Socialization is never total; it is never complete. The subordinate institutional orders represent weak points in a system of social control—actual and potential points where intervention to usher in a more humane society is indeed possible.

The Family

Given its role in the socialization of the young, the family is crucial to the establishment of legitimacy. Therefore, it is also crucial for challenging this legitimacy. The family has tremendous potential for the instillation of autonomy and self-worth in its members. A basic tenet of humanist sociology is that hegemonic domination is never complete even in the most totalitarian system. Complete control of the family is more difficult than control of other social institutions, because the family is not a centralized unit. The right of parents to raise their children is deeply embedded in our society and is legitimized. Sociological research shows us that the various classes raise their children differently. Working-class parents bring up their children to follow the orders of authorities (parents, teachers, police officers, bosses), but middle-class parents encourage their children to question and to seek autonomy.[9] A traditional sociologist would stop at reporting this finding, but a humanist sociologist would use this information to point out that this should and could be changed. Those lower- and working-class parents who are raising their children to follow orders must be informed by humanist sociologists that they can raise their offspring in a more open manner, that obedience and acquiescence to authority is not the only way to raise children. It should be stressed, too, that parents should only be made aware of this, not forced to act on it. Given the opportunity, it is almost certain they would choose freedom and autonomy for themselves and their children.

Some radical feminists have been quick to call for the abolition of the family, seeing it as an instrument of oppression. But no one has argued convincingly that families cannot provide emotional stability and strength, only that some families may not have done so. Therefore, what is needed is a plan that encourages the family to instill humane values rather than the materialistic, consumer-oriented values.

There are signs that family values can and are changing. Foremost among the forces for change is the women's movement. The women's movement came into being at a time when more women were entering the paid economy (the mid to late 1960s), and since then has contributed to the greater influence of women in the work force. Projections by the U.S. Department of Labor predict that in 1995 women will hold 54 percent of all jobs[10] This indicates that women are gaining a measure of economic self-sufficiency, which in turn is producing changes in the division of labor within the American family. Although men, even though they are on the whole not carrying a proportional role in what had previously been thought of as women's work in the home—helping with cooking, cleaning, and the socialization of children—have made some progress in this direction.[11] Thus, it is becoming more and more clear that the women's movement is about more than just women getting jobs. As more women enter the labor force and share in the breadwinning, family bonds and values seem to strengthen. Strong emotional bonds are being established between husbands and wives, parents and children.

Through the efforts of the women's movement, a condition in which the family, usually a bastion of conservatism, instead is becoming a progressive political and social force. Betty Friedan, who is thought of as the "mother" of the women's liberation movement, rhetorically asks:

> For when men start assigning a higher priority to their families and self-fulfillment, and women a higher priority to independence and active participation in "man's world," what happens to the supremacy of the corporate bureaucratic system?[12]

Restrictive options have provided a destructive milieu for women. The fact that the women's movement has challenged these restrictions for both women and men shows that the status quo can indeed be shaken. Women have questioned the legitimacy of men who are in positions of power and who previously had defined their lives for them. A case in point was the recent U.S. Senate Judicial Committee Hearings on the nomination of Clarence Thomas for Associate Justice of the Supreme Court. Although Judge Thomas did win approval, the accusations of sexual harassment by Anita Hill not only changed the way this nation looks at sexual harassment but also was one of the reasons that 1992 saw the election of four more women to the U.S. Senate and a fifth in a 1993 special election, whereas there had only been two prior to the Thomas/Hill hearings. Women around the country began to question the legitimacy of the dominant group, thereby affecting change.

In short, although the women's movement has been primarily supported by middle-class college-educated women and men, it offers potential for changing the whole of society. In particular, it has alerted women to the institutional nature of their oppression. They have sought to reconstruct social institutions, and the changes in the institution of the family are already beginning to affect the total society. Whether or not American society can accommodate a strengthened family is something that can only be answered in the future. However, the fact that a segment of society has changed (no matter how little) is proof that change is indeed possible. It is upon this potential for restructuring the basic institution of the family that a humanistic society can be built. But the family cannot accomplish this alone. The other institutional orders also must play a role in changing American society.

Religion

Religion is a prime source of self-esteem and meaning. Here Ernest Becker's insight into the potential of religion for a humanistic society should be noted. According to Becker, death is the primary repression, and human beings are creatures who strive for immortality. Society is built upon the repression of the thought of death and the desire for immortality. Thus society enables human beings to meet the terror of life and loneliness by identifying with the society as a vehicle for immortality. Power also is related to the repression of thoughts of death. Individuals have always been fascinated with those who are in positions of power, and as members of groups they can identify with a leader who enables them to overcome the loneliness and the smallness they feel. This identification process also explains why society rushes to name airports, stadiums, and streets after dead heroes. It is as though we want "to declare that he will be immortalized physically in the society, in spite of his physical death."[13]

When the leader dies, it is a stark reminder to people of the terror of their own death. If one who was so powerful can be brought down, what chance do ordinary mortals have? And when the slain leader is young and vital, as was John Kennedy, the starkness of our own mortality is even more apparent. Did those individuals who wept for President Kennedy weep for him or for themselves? Becker's answers to these questions are:

> We don't know, on this planet, what the universe wants from us or is prepared to give us. We don't have an answer to the question that troubled Kant of what our duty is, what we should be doing on earth. We live in utter darkness about who we are and why we are here, yet we know it must have some meaning. What is more natural, then, than to take this unspeakable mystery and dispel it straight-away by addressing our performance of heroics to another human being, knowing thus daily whether the performance is good enough to earn us immortality.[14]

Human beings have a need for hope and for meaning, and these needs make us reach out for something that makes life worthwhile. Becker offers four levels of meaning by which an individual can choose to live:

1. The first, most intimate, basic level is the personal one. It is the level of one's "true" self, one's special gift or talent, what one feels deep down inside, the person one talks to when alone, the secret self of one's inner scenario.
2. The second, or next highest level is the social. It represents the most intimate extension of oneself to a select few intimate others: spouse, friends, relatives, perhaps even pets.
3. The third and next highest level is the secular. It consists of symbols of allegiance at a greater personal distance and often higher in power and compellingness: the corporation, the party, the nation, science, history, humanity.
4. The fourth and highest level of power and meaning is the sacred: It is the invisible and unknown level of power, the insides of nature, the source of creation, God.[15]

Becker chose the fourth level, claiming that true meaning could only be cosmic. To Becker, what humankind feared most was not extinction itself, but extinction without meaning. The self-perpetuation motivation of individuals ultimately transfers to the religious sphere. This is the crux of renowned psychiatrist Otto Rank's critique of psychology as self-deception.[16] All that psychology—and by extension, social science—has been able to accomplish in its efforts to become a "science" is the erosion of the belief that individuals have a soul. All that social science provides is a form of reflexivity that inevitably leads to isolation. In Becker' thought, only religion can overcome the crippling fear of death.

Most sociologists—including humanists—may not agree with Becker's attitude toward religion. Whether Becker was right or wrong in his view that religion provides the ultimate meaning for human beings is inconsequential, given his insight that individuals identify with such external factors as organizations and leaders in order to gain a measure of self-esteem. If, oppression is institutionalized and supported by those in power, then this identification is misguided. The religious institutional order has a different base from the materialism of political and economic structures. Because of an emphasis upon the meaning of life, there is almost automatic competition between Church and State for people's loyalty, attention, commitment, and time.

Two examples of this potential conflict stand out: the participation of clergy in both the Civil Rights movement in the United States and the Theology of Liberation movement in Latin America. Although the hierarchies of the churches in the United States initially did little to support the Civil Rights movement, individual clergy were a significant force in the movement. Protestant, Catholic, and Jewish clergy participated from the first in Civil Rights marches, sit-ins, and demonstrations.

Martin Luther King, Jr., whose nonviolence was rooted in his religious training, would not have attracted the early following he did had he not been a minister. The role of religious individuals in the Civil Rights movement has been described by Ronald Johnstone:

> Martin Luther King, Jr., challenged church people of all colors to support the struggle of black citizens for freedom and dignity. Seminaries intensified their efforts by offering new courses in social ethics and concerns and by implementing social action field-work programs. Denominations began issuing position statements supporting, at least in general terms, the rights of minority and disprivileged groups. Young seminary graduates assuming the leadership of congregations began preaching a gospel that included more than personal comfort and salvation.[17]

There is little doubt that the Civil Rights movement would not have accomplished as much as it did without the dedicated clergy of all faiths and colors, who, motivated by a belief in the equality of all individuals before God, put their beliefs into practice.

The second example, the Theology of Liberation, which is quite prominent in Latin America, also attests to the radical potential of religion. Throughout Latin America priests not only are preaching human dignity and social reform but also are actively involved in political organization.[18] The image of the "guerrilla-priest" has spread throughout Latin America. Priests seek to overcome the resistance of the Latin American elites to changing the status quo, a continuing state that fosters poverty and misery for the masses of Latin American peoples.

Religion thus holds more than a potential for change. In these examples, as well as in countless others, clergy and those inspired by religious beliefs have tried to institute procedures for establishing humanistic values of equality, justice, and freedom, and have succeeded. Religion does not have to be Marx's "opium of the people"; it can be in the vanguard for change. Organized religion, which in the past has tended to support oppressive regimes can, if it is pushed by individuals within its structure, withdraw its legitimacy, because it receives its authority from an entirely different source than does the government. Power structures that do not have the support of religion could quickly find themselves in a situation of conflict. However, although religion holds a powerful potential for change, it also (as seen in chapter 7) holds a powerful potential for repression. This has been shown by the example of the "Religious Right" in America, who have sought to impose their own conservative agenda on the rest of the nation.

Education

Schooling provides an important source of legitimation for the society. But, as with the family and religion, there are alternatives to this position. A prime source for change is in higher education, where academic freedom and tenure have been institutionalized.

The history of higher education has been one of constant conflict, with legitimacy exceedingly difficult to establish and maintain. At first the conflict centered around the establishment of stable institutions of learning in a state of what one writer called "happy anarchy,"[19] in which anyone with the necessary funds (individuals, private groups, religious bodies, units of government) could and did establish colleges and universities.

As some colleges developed into universities, strife broke out between the various factions within the institution. Thus a tradition of conflict (which, by definition, raises the specter of illegitimacy) was established. One of the most cherished and hard-won battles occurred over the professionalism of faculty and their desire to gain the rights of tenure and academic freedom.

Professors who espouse unpopular ideas must be protected. Professors theoretically are bound to pursue truth wherever it may lead, and this can produce opposition with other segments of the population who do not hold these lofty ideals. Although the need for academic freedom grew out of the controversy over Darwinism and the teaching of principles of evolution, it eventually came to incorporate political ideas. Friction was created when academicians sought to assimilate the doctrine of free speech into that of academic freedom. Social scientists in particular came into conflict with important business interests in the world outside the university. Historian Frederick Rudolf has catalogued some of the more famous of these conflicts that arose in the late nineteenth century.

> Richard T. Ely, economist at the University of Wisconsin, spoke favorably of strikes and boycotts; for his economic heresy he was tried by a committee of the Board of Regents in 1894. Edward W. Bemis, economist at the University of Chicago, chose the Pullman strike to make a public attack on the railroads; he was dismissed. James Allen Smith, political scientist at Marietta College, did not like monopolies, and he met the same fate as Bemis. President E. Benjamin Andrews of Brown revealed a preference for free trade and bimetalism that made his position at an old eastern college untenable. John R. Commons indulged in a range of economic views so disturbing that he ran into difficulty at both Indiana University and Syracuse. Edward A. Ross, sociologist at Stanford, disapproved of coolie labor, with the consequence that Mrs. Leland Stanford disapproved of him.[20]

From these early skirmishes, which were fought for academic freedom and which professors usually lost, emerged a determined professorate, which eventually won the war.

In the early decades of the twentieth century the principle of academic freedom became established as a professional prerogative and evolved into a safeguard for both professors' freedom of thought and their economic security. Today, the principles of academic freedom and tenure are firmly entrenched, and professors, more than most other groups in this society, enjoy a great deal of freedom. However, this is not the case for their colleagues who teach at the elementary and secondary levels. Those professors who believe in humanitarian values must not only teach and practice these values but also work to see that elementary and secondary teachers enjoy the same rights and privileges. Our society has accepted the principle that educational credentials are exceedingly important. (One must look hard and far to find a political or economic leader who is not at least a college graduate.) Those individuals who eventually fill the positions in the power structure must pass through the educational institutional order. Thus, the opportunity for restructuring education to provide a humanist-based learning should not be wasted. Tenured professors cannot be fired for teaching and practicing the values of equality, justice, and fairness. Humanist sociologists must join with humanists in other disciplines and in other levels of education in order to foster a more humane society. By pointing out to students society's constraints and controls and how these can be changed, education offers real hope for structural changes in the society.

Again, as with religion, a note of caution must be sounded. Because of unpopular stances, many professors never achieve the security of tenure and academic freedom. In an era of fiscal constraint, where jobs are difficult to come by, a chilling effect may push young academics to silence. This must be rigorously guarded against. Furthermore, the United States is witnessing an unprecedented assault on the elementary and secondary system of education by religious fundamentalists who are seeking to gain control over local school boards and thereby control education at these levels. It therefore is even more imperative for humanist sociologists to speak up and to act to offset these dangerous attempts at ideological control.

The Mass Media

As we saw in chapter 7, both advertising and television play a large role in establishing the hegemonic domination of the power structure. Given the extent of corporate control over television programs (advertisers are not going to sponsor shows that are critical of the status quo) there is little hope for change occurring here. The news media, on the other hand, with their belief in the ideal of the objective pursuit of truth and the recent success of some investigative reporters, do present a potential for change. By presenting accurate information to the people, the news media can play a role in questioning the doings of the power structure. Herbert Gans offers us a way in which the public can be informed—through what he calls "multiperspectival news."[21]

Multiperspectival news differs from traditional news reporting in five ways:

1. Multiperspectival news is more national. It goes beyond the current equating of the federal government with the nation and seeks to report comprehensively about more national and nationwide agencies and institutions. These would include national corporations, unions, and voluntary associations, as well as organized and unorganized interest groups.

2. Multiperspectival news would add a bottom-up view to the current top-down approach. Individuals and groups usually excluded from the news because of their powerlessness would be given courage.

3. Multiperspectival news would feature more output news, determining how the plans and programs of national and nationwide public and private agencies have worked out in practice for intended and unintended beneficiaries, victims, bystanders, and the general public.

4. Multiperspectival news would aim to be more representative, reporting on the activities and opinions of ordinary Americans from all population sectors and roles. Representative news would mean greater coverage of the diversity of opinion, from many positions in different national and social hierarchies, and from many points on the ideological spectrum.

5. Multiperspectival news would place more emphasis on service news, providing relevant information for specific national sectors and roles: what people consider to be important national news about themselves.[22]

The fundamental justification for multiperspectival news is its potential for furthering a humanistic society. Although it is naive to assume that increased knowledge will by itself redistribute power, nevertheless, the additional knowledge along with change in the other subordinate institutional orders may do just that. Newscasters and journalists are in a pivotal position in American society. Although journalists, as we saw in chapter 7, are themselves basically upper-middle class and carry with them middle-class values, they do hold out some hope for affecting change. Journalists are, by their position in society, training, and education, professionals. With professionalism goes autonomy. A handful of reporters and journalists, by pursuing such stories as My Lai, the Pentagon Papers, Watergate, Irangate, and the Savings and Loan scandals, to mention only a few major ones, have, if not solidified, at least established precedents for themselves and for their colleagues around the country for a degree of freedom of the press. Freedom once given is extremely difficult to take away. A democracy can be supported only by an informed citizenry. If newscasters can begin to ask the right questions and report answers in the form of alternatives, freedom of the press may translate into freedom of the people.

Humanist Society or Utopian Dream?

Is a humanistic society really possible or is it another utopian dream? The answer is yes, it is possible.

But in order to begin the reconstruction of the restrictive society that dominates our lives humanist sociology must be taken seriously. Humanist sociologists must disseminate to the general public the knowledge and research they have accumulated. Humanist sociologists should offer the following ten basic tenets of humanist sociology, all of which have been presented in some form in this text, to the American people.

1. *Human beings are free reflexive individuals who can choose between alternatives.* If individuals come to think of themselves as free and believe that they have free choice; they will not readily follow all orders. They will not easily be controlled.
2. *Roles must be looked at reflexively.* On the one hand roles enable us to develop a sense of self, but equally important is that roles tend to restrict human beings when they are defined in narrow ways.
3. *The advocacy of reflexivity states that consciousness can be changed.* The women's movement, for example, has "raised" people's consciousness.
4. *Self-esteem and its maintenance are of paramount importance to human beings.* Steps must be taken to insure that individuals are allowed to establish self-esteem, because human beings who feel good about themselves have little reason to wreak havoc on others.
5. *By uncritically accepting the ideology of individual achievement and responsibility, individuals do not look to the structure of the society as a locus for their problems.* By turning inward and blaming themselves for their failure, individuals will not seek to change the conditions under which they live.
6. *Professionalism carries with it autonomy.* Professionals, by embracing autonomy as part of their occupational values, can become a powerful force for change. Journalists who seek to report the "truth" will overcome obstacles to do so. College professors, in

particular, have the freedom to try to construct ways in which we can achieve equality, justice, and liberation.

7. *Social structures define our existence.* Points of intervention in the social structure must be located if change is desired.

8. *Elites of power are not omnipotent.* They make mistakes. They have clashes among themselves. Hegemonic domination is never complete. The socialization process is too varied and diverse to be completely controlled.

9. *Conflict is endemic to society.* Conflict is always at the surface or just below it in any society. Conflict is a threat to legitimation.

10. *Legitimacy holds the society together.* Dominant institutional orders that have their legitimacy withdrawn cannot survive.

Should a humanist sociology be taken seriously by the general public, then a society based on justice, equal opportunity, freedom from undue restrictions, and similar values could begin to be built. Now is the time for those who believe in these principles to try to restructure American society. Although I would like to say that such an endeavor will be successful, there can be no guarantees. The only guarantee that can be given is that if no attempt is made, there will be more of the same. It is now up to us to either build a new path or continue on the narrow road that others have built for us.

Summary and Conclusion

Although this text is critical of contemporary American society, it should be clear that America is not a totalitarian state, nor will it become one in the near future. While it is easy to concentrate on the repressive tendencies of the United States, a more realistic analysis would show that America has great potential for freedom. A small group of individuals may be trying to control people's lives, but this control is never complete. Neither has this group usurped the democratic ideals upon which this nation was founded. Americans believe in freedom, justice, fairness, equality, and other humanistic values. Moreover, these values are professed publicly and are guaranteed by the Constitution, the Bill of Rights, and the legal system. Although the rich and powerful often unfairly use legal means to protect their interests and subsequently are treated more favorably in the legal system than are the poor and the powerless,[23] the rights of the poor have not been completely stamped out. This is important to keep in mind because these rights can be used by the powerless once they know how to do so.

Humanist sociology holds that if people are given accurate information upon which to make judgments and are provided the opportunity upon which to make judgments and are provided the opportunity to carry them out, the overwhelming majority of people would choose humane values. People would want to eliminate poverty, would feel that racism and sexism are evil, would hold that people are more important than property, would choose freedom over control.

Humanist sociology is by definition antiestablishment. It opts for intervention and change at the expense of the status quo. As such it comes into direct conflict with the power structure. In the words of Alfred McClung Lee, a co-founder of the Association for Humanist Sociology and, until his death in 1992, the most well-known and influential humanist sociologist in the world, humanist sociology

> *can* show people how the control procedures of the family and other social agencies tend to push most of us ... toward accepting roles as willing and dependent instruments for others. It *can* substitute an accurate conception for an unthinking or sentimental conception of social controls. It *can* demonstrate how the control of power in our society depends more on loyalty than upon competence, and therein lies the controllers' greatest liability. It *can* spell out the insecurity and frailty of anyone controlling social power....[24]

Notes

1. Ernest Becker, *The Birth and Death of Meaning*, 2d ed. (New York: Braziller, 1971), p. 34.

2. Ibid., p. 62.

3. Ibid., p. 67.

4. Ibid., p. 191.

5. Ernest Becker, *Revolution in Psychiatry* (New York: Free Press, 1974), p. 219.

6. C. Wright Mills, *The Sociological Imagination* (New York: Oxford University Press, 1959).

7. Ernest Becker, *The Structure of Evil* (New York: Braziller, 1968), p. 163.

8. Becker, *Birth and Death of Meaning*, p. 177.

9. Melvin Kohn, *Class and Conformity* (Homewood, IL: Dorsey, 1969); "Social Class and the Exercise of Parental Authority," *American Sociological Review* 24(1959):352–66; "Social Class and Parental Values: Another Confirmation of the Relationship," *American Sociological Review* 61(1976):538–65, 568; and Melvin Kohn and Carmi Schooler, "Job Conditions and Personality: A Longitudinal Assessment of their Reciprocal Effects," *American Journal of Sociology* 87(1982):1257–86.

10. U.S. Department of Labor. *Monthly Labor Report, June.* (Washington, DC: U.S. Government Printing Office, 1992).

11. John P. Robinson, "Who's Doing the Housework?" *American Demographics* 10(1988):24–28; Myra Max Ferree, "Feminism and Family Research,"

in Alan Booth, ed., *Contemporary Families* (Washington, DC: National Council on Family Relations, 1991), pp. 103–21.

12. Betty Friedan, "Feminism Takes a New Turn," *New York Times Magazine*, Nov. 18, 1979, p. 100.

13. Ernest Becker, *The Denial of Death* (New York: Free Press, 1973), p. 149.

14. Ibid., p. 156.

15. Becker, *Birth and Death of Meaning*, p. 186.

16. See Otto Rank, *Psychology and the Soul* (New York: Perpetua Books, 1961).

17. Ronald L. Johnstone, *Religion and Society in Interaction* (Englewood Cliffs, NJ: Prentice-Hall, 1975), p. 20.

18. See Roger Lancaster, *Praise God and the Revolution* (Berkeley: University of California Press, 1990).

19. Paul Woodring, *The Higher Learning in America: A Reassessment* (New York: McGraw-Hill, 1968), p. 3.

20. Frederick Rudolf, *The American College and University: A History* (New York: Vintage, 1965), p. 414.

21. Herbert J. Gans, *Deciding What's News* (New York: Pantheon, 1979), pp. 313–22.

22. Adapted from ibid., pp. 313–14.

23. See Jeffrey Reiman, *The Rich Get Richer and the Poor Get Jail* (New York: Macmillan, 1984).

24. Alfred McClung Lee, *Toward Humanist Sociology* (Englewood Cliffs, NJ: Prentice-Hall, 1973), p. 36.

BIBLIOGRAPHY

Alexander, Jeffrey. "Neo-Functionalism Today: Reconstructing a Theoretical Tradition." In George Ritzer, ed., *Frontiers of Social Theory*. New York: Columbia University press, 1990.

———. "Toward Neo-Functionalism." *Sociological Theory* 3 (1985): 11–23.

———. *Theoretical Logic in Sociology*, vol. 4: *The Modern Reconstruction of Classical Thought*. Berkeley: University of California Press, 1983.

Allen, Michael Patrick. "Continuity and Change within the Core Corporate Elite." *Sociological Quarterly* 19 (1978): 510–21.

Anderson, Charles H. *The Political Economy of Social Class*. Englewood Cliffs, NJ: Prentice-Hall, 1974.

———. *Toward a New Sociology*. Rev. ed. Homewood, IL: Dorsey, 1974.

Anderson, Nels. *The American Hobo: An Autobiography*. 1923. Leiden, The Netherlands: Brill, 1975.

Antonio, Robert J. "The Problems of Normative Foundations in Emancipatory Theory: Evolutionary versus Pragmatic Perspectives." *American Journal of Sociology* 94 (1989): 721–48.

Atkinson, Dick. *Orthodox Consensus and Radical Alternative*. New York: Basic Books, 1972.

Auerbach, Stuart. "U.S. Relied on Foreign-Made Parts for Weapons." In Hedrick Smith, ed., *The Media and the Gulf War*. Washington, DC: Seven Locks Press, 1992.

Barnett, Richard, and Ronald E. Muller. *Global Reach: The Power of the Multinational Corporation*. New York: Simon and Schuster, 1974.

Baudrillard, Jean. *In the Shadow of the Silent Minorities*. New York: Semiotext, 1983.

Becker, Ernest. *Revolution in Psychiatry*. New York: Free Press, 1974.

———. *The Denial of Death*. New York: Free Press, 1973.

_____. *The Birth and Death of Meaning.* 2d ed. New York: Free Press, 1971.

_____. *The Lost Science of Man.* New York: Braziller, 1971.

_____. *The Structure of Evil.* New York: Braziller, 1968.

_____. *Beyond Alienation.* New York: Braziller, 1967.

Bennett, W. Lance. *The Governing Crisis: Media, Money, and Marketing in American Elections.* New York: St. Martin's Press, 1992.

Bensman, Joseph, and Arthur J. Vidich. *The New American Society: The Revolution of the Middle Class.* Chicago, IL: Quadrangle, 1971.

Berger, Peter L. *Invitation to Sociology: A Humanistic Perspective.* Garden City, NY: Anchor, 1963.

Berger, Peter L., and Thomas Luckmann. *The Social Construction of Reality.* Garden City, NY: Doubleday, 1966.

Berle, A. A., Jr., and Gardiner Means. *The Modern Corporation and Private Property.* New York: Macmillan, 1933.

Bernstein, Richard J. *The Reconstruction of Social and Political Theory.* Philadelphia: University of Pennsylvania Press, 1978.

Biddle, Bruce J. and Edwin J. Thomas, eds. *Role Theory: Concepts and Research.* New York: Wiley, 1963.

Bierstedt, Robert. *Power and Progress: Essays on Sociological Theory.* New York: McGraw-Hill, 1975.

_____. *The Social Order.* 3d ed. New York: McGraw-Hill, 1970.

Blau, Peter M., ed. *Approaches to the Study of Social Structure.* New York: Free Press, 1975.

_____. *The Dynamics of Bureaucracy.* 2d rev. ed. Chicago, IL: University of Chicago Press, 1973.

Blau, Peter M., and Otis Dudley Duncan. *The American Occupational Structure.* New York: Free Press, 1978.

Blumer, Herbert. *Symbolic Interactionism.* Englewood Cliffs, NJ: Prentice-Hall, 1986.

_____. "Society and Symbolic Interactionism." In Arnold Rose, ed., *Human Behavior and Social Process.* Boston, MA: Houghton Mifflin, 1962.

Board of Governors of the Federal Reserve System. *Federal Research Bulletin.* Washington, DC: U.S. Government Printing Office, 1992.

Bogdan, Robert, and Stephen J. Taylor. *Introduction to Qualitative Research Methods.* New York: Wiley, 1975.

Bollinger, Lee C. *Images of a Free Press.* Chicago, IL: University of Chicago Press, 1991.

Bottomore, Thomas B. "Marxist Sociology." In David Sills, ed., *The International Encyclopedia of the Social Sciences,* vol. 10. New York: Macmillan/Free Press, 1968.

Braginsky, E. F., and Braginsky, D. D. *Mainstream Psychology: A Critique.* New York: Holt, Rinehart and Winston, 1974.

Breland, Keller, and Marian Breland. "The Misbehavior of Organisms." *American Psychologist* 16 (1961): 681–84.

Brim, Orville, G., and Stanton Wheeler. *Socialization after Childhood: Two Essays.* New York: Wiley, 1966.

Bruyn, Severyn T. *The Human Perspective in Sociology*. Englewood Cliffs, NJ: Prentice-Hall, 1966.

Burch, Philip H., Jr. *The Managerial Revolution Reassessed*. Lexington, MA: D.C. Heath, 1972.

Capps, Walter H. *The New Religious Right: Piety, Patriotism and Politics*. Columbia: University of South Carolina Press, 1990.

Chafetz, Janet. *Feminist Sociology*. Itasca, IL: Peacock, 1988.

Chambliss, William, ed. *Sociological Readings in a Conflict Perspective*. Reading, MA: Addison-Wesley, 1973.

Chandler, Alfred. *The Visible Hand: The Managerial Revolution in American Business*. Cambridge, MA: Harvard University Press, 1977.

Charnofsky, Stanley. *Educating the Powerless*. Belmont, CA: Wadsworth, 1971.

Charon, Joel M. *Symbolic Interactionism*. Englewood Cliffs, NJ: Prentice-Hall, 1979.

Clinard, Marshall B. "The Sociologists' Quest for Respectability." *Sociological Quarterly* 7 (1966): 399-412.

Coleman, James S., Ernest Q. Campbell, Carol J. Hobson, James McPortland, Alexander M. Mood, Frederick D. Weinfield, and Robert L. York. *Equality of Educational Opportunity*. 2 vols. Washington, DC: U.S. Government Printing Office, 1966.

Coleman, James. ed. *Education and Political Development*. Princeton, NJ: Princeton University Press, 1965.

Coll, Steven, and William Branigan. "U.S. Scrambled to Shape Highway of Death." In Hedrick Smith, ed., *The Media and the Gulf War*. Washington, DC: Seven Locks Press, 1992.

Collins, Randall. "Conflict Theory and the Advance of Macro-Historical Sociology." In George Ritzer, ed., *Frontiers of Social Theory*. New York: Columbia University Press, 1990.

———. *Theoretical Sociology*. San Diego, CA: Harcourt Brace Jovanovich, 1988.

———. *Conflict Sociology: Toward an Explanatory Science*. New York: Academic Press, 1975.

———. "Functional and Conflict Theories of Educational Stratification." *American Sociological Review* 36 (1971): 1002-18.

Colomy, Paul. "Recent Developments in the Functionalist Approach to Change." *Sociological Focus* 51 (1986): 139-58.

Comte, Auguste. *The Positive Philosophy of Auguste Comte*, Harriet Martineau, trans. and ed. London: J. Chapman, 1853/1896.

Conwell, Chic, and Edwin H. Sutherland. *The Professional Thief*. Chicago, IL: University of Chicago Press, 1937.

Cooley, Charles Horton. *Social Organization*. New York: Schocken, 1962.

Coser, Lewis A. *Masters of Sociological Thought*. 2d ed. New York: Harcourt Brace Jovanovich, 1977.

Coser, Rose L. *In Defense of Modernity*. Stanford, CA: Stanford University Press, 1991.

Curtiss, Susan. *Genie: A Psychological Study of a Modern Day "Wild Child."* New York: Academic Press, 1977.

Dahl, Robert A. *Who Governs? Democracy and Power in an American City.* New Haven, CT: Yale University Press, 1961.

———. *A Preface to Democratic Theory.* Chicago: University of Chicago Press, 1956.

Dahrendorf, Ralf. *Essays in the Theory of Society.* Stanford, CA: Stanford University Press, 1968.

———. *Class and Class Conflict in Industrial Society.* Stanford, CA: Stanford University Press, 1959.

Davis Kingsley. "Final Note on a Case of Extreme Isolation." *American Journal of Sociology* 52 (1947): 432–37.

———. "Extreme Social Isolation of a Child." *American Journal of Sociology* 45 (1940): 554–64.

Davis, Kingsley, and Wilbert E. Moore. "Some Principles of Stratification." *American Sociological Review* 10 (1945): 242–49.

Dawson, Richard E., Kenneth Prewitt, and Karen S. Dawson. *Political Socialization.* 2d ed. Boston, MA: Little, Brown, 1977.

Delfattore, Joan. *What Johnny Shouldn't Read.* New Haven, CT: Yale University Press, 1992.

Demott, Benjamin. *The Imperial Middle: Why Americans Can't Think Straight about Class.* New York: Morrow, 1990.

Denzin, Norman K. *The Research Act,* 3d ed. New York: McGraw-Hill, 1989.

Deutscher, Irwin. "Words and Deeds: Social Science and Social Policy." *Social Problems* 13 (1965): 233–54.

Dollard, John. *Criteria for the Life History.* Freeport, NY: Books for Library Press, 1971.

Domhoff, G. William. *The Power Elite and the State: How Policy Is Made in America.* New York: Aldine De Gruyter, 1990.

———. *Who Rules America Now?* Englewood Cliffs, NJ: Prentice-Hall, 1983.

———. *The Powers That Be: State and Ruling Class in Corporate America.* New York: Random House, 1979.

———. *Who Really Rules? New Haven and Community Power Reexamined.* Santa Monica, CA: Goodyear, 1978.

———. *The Bohemian Grove and Other Retreats: A Study in Ruling Class Cohesiveness.* New York: Harper and Row, 1975.

———. *The Higher Circles.* New York: Random House, 1970.

———. *Who Rules America?* Englewood Cliffs, NJ: Prentice-Hall, 1967.

Domhoff, G. William, and Hoyt B. Ballard, eds. *C. Wright Mills and the Power Elite.* Boston, MA: Beacon Press, 1968.

Duke, James K. *Power and Conflict in Social Life.* Provo, UT: Brigham Young University Press, 1976.

Duncan, Otis Dudley. "Methodological Issues in the Analysis of Social Mobility." In Neil J. Smelser and Seymour Martin Lipset, eds., *Social Structure and Mobility in Economic Development.* Chicago, IL: Aldine, 1966.

Durkheim, Emile. *Suicide.* 1897. Glencoe, IL: Free Press, 1951.

———. *The Rules of Sociological Method.* 1895. Glencoe, IL: Free Press, 1950.

———. *The Division of Labor in Society.* 1893. Glencoe, IL: Free Press, 1956.

Dye, Thomas R. *Who's Running America?* Englewood Cliffs, NJ: Prentice-Hall, 1986.

Edwards, Harry. "The Black Athletes: Twentieth Century Gladiators for White America." *Psychology Today* 7 (Nov. 1973): 40–52.

Ericson, Kai T. "A Comment on Disguised Observation in Sociology." In William J. Filstead, ed., *Qualitative Methods: Firsthand Involvement with the Social World.* Chicago, IL: Markham, 1970.

Ewen, Stuart. *Captains of Consciousness.* New York: McGraw-Hill, 1976.

Featherman, David L. *Has Opportunity Declined in America?* Institute for Research on Poverty, Discussion Paper No. 437–77. Madison: University of Wisconsin, 1977.

Ferree, Myra Max. "Feminism and Family Research." In Alan Booth, ed., *Contemporary Families.* Washington, DC: National Council on Family Relations, 1991.

Festinger, Leon, Henry Riecken, and Stanley Schacter. *When Prophecy Fails.* New York: Harper and Row, 1956.

Filmer, Paul, Michael Phillipson, David Silverman, and David Walsh. *New Directions in Sociological Theory.* Cambridge, MA: MIT Press, 1973.

Filstead, William J., ed. *Qualitative Methodology: Firsthand Involvement with the Social World.* Chicago, IL: Markham, 1970.

Freitag, Peter. "The Cabinet and Big Business: A Study of Interlocks." *Social Problems* 23 (1975): 137–52.

Freud, Sigmund. *New Introductory Lectures on Psychoanalysis,* James Strachey, ed. and trans. New York: Norton, 1965.

———. *Civilization and Its Discontents.* James Strachey, ed. and trans. New York: Norton, 1962.

Freund, Julien. *The Sociology of Max Weber.* New York: Pantheon, 1968.

Friedan, Betty. "Feminism Takes a New Turn." *New York Times Magazine,* Nov. 18, 1979, pp. 40, 92–106.

Galbraith, John Kenneth. *The New Industrial State.* New York: Signet, 1968.

Gans, Herbert J. *Deciding What's News.* New York: Pantheon, 1979.

Garfinkle, Harold. *Studies in Ethnomethodology.* Englewood Cliffs, NJ: Prentice-Hall, 1967.

Gay, Craig M. *With Liberty and Justice for Whom? The Recent Evangelical Debate over Capitalism.* Grand Rapids, MI: Eerdmans, 1991.

Gergen, Kenneth. *The Saturated Self.* New York: Basic Books, 1991.

Gerth, H. H. and C. Wright Mills. *Character and Social Structure.* New York: Harbinger, 1964.

Gerth, Hans, and C. Wright Mills, eds. and trans. *From Max Weber.* New York: Oxford University Press, 1946.

Giddens, Anthony. *The Constitution of Society.* New York: Columbia University Press, 1984.

Gilbert, Dennis, and Joseph Kahl. *The American Class Structure.* Homewood, IL: Dorsey, 1987.

Gilligan, Carol. *In a Different Voice.* Cambridge, MA: Harvard University Press, 1982.

Gittlin, Todd. "Prime Time Ideology: The Hegemonic Process in Television Entertainment." *Social Problems* 26 (1979): 251–66.

Glass, John F. "The Humanistic Challenge to Sociology." *Journal of Humanistic Psychology* 11 (1971): 170–83.

Goffman, Erving. *Gender Advertisements.* New York: Harper and Row, 1979.

———. *Frame Analysis.* New York: Harper and Row, 1974.

———. *Behavior in Public Places: Notes on the Social Organization of Gatherings.* New York: Free Press, 1963.

———. *Encounters: Two Studies in the Sociology of Interaction.* Indianapolis, IN: Bobbs-Merrill, 1961.

———. *The Presentation of Self in Everyday Life.* Garden City, NY: Anchor, 1959.

Goldsmith, William S., and Edward J. Blakely. *Separate Societies: Poverty and Inequality in U.S. Cities.* Philadelphia, PA: Temple University Press, 1992.

Goode, William J. "Norm Commitment and Conformity to Role-Status Obligations." In Bruce J. Biddle and Edwin J. Thomas, eds., *Role Theory: Concepts and Research.* New York: Wiley, 1963.

Gouldner, Alvin W. *The Coming Crisis of Western Sociology.* New York: Basic Books, 1970.

Greenberg, Bradley S., Marcia Richardson, and Laura Henderson," Trends in Sex-Role Portrayals on Television." In Bradley S. Greenberg, ed., *Life on Television.* Norwood, NJ: Ablex, 1980.

Gugliotta, Guy. "Number of Poor Americans Rises for 3rd Year." *Washington Post,* Oct. 5, 1993, p. A6.

Habermas, Jurgen. *The Theory of Communicative Action.* Vol. 2. Boston, MA: Beacon Press, 1987.

———. *The Theory of Communicative Action.* Vol. 1. Boston, MA: Beacon Press, 1984.

———. *Communication and the Evolution of Society.* Boston, MA: Beacon Press, 1979.

———. *Legitimation Crisis.* Boston, MA: Beacon Press, 1975.

———. *Knowledge and Human Interests.* Boston, MA: Beacon Press, 1971.

———. *Toward a Rational Society.* Boston, MA: Beacon Press, 1970.

Halberstam, David. *The Powers That Be.* New York: Knopf, 1979.

Hall, Calvin S., and Gardiner Lindzey. *Theories of Personality.* New York: Wiley, 1966.

Hall, Peter. "A Symbolic Interactionist Analysis of Politics." *Sociological Inquiry* 42 (1972): 35–75.

Hebding, Daniel E. and Leonard Glick. *Introduction to Sociology.* Reading, MA: Addison-Wesley, 1981.

Hilbert, Richard A. "Toward an Improved Understanding of 'Role.'" *Theory and Society* 10 (1981): 207–22.

Hills, Stuart, and Ron Santiago. *Tragic Magic.* Chicago, IL: Nelson-Hall, 1992.

Hodge, Robert W., Paul M. Siegel, and Peter H. Rossi. "Occupational Prestige in the United States: 1925–1963." *American Journal of Sociology* 70 (1964): 286–302.

Hollingshead, August, and Frederick Redlich. *Social Class and Mental Illness: A Community Study.* New York: Wiley, 1958.

Horowitz, Irving Louis, and Lee Rainwater. "Journalistic Moralizers." *Transaction* 7 (1970): 5–8.

Humphreys, Laud. *Tearoom Trade*. Chicago: Aldine, 1970.

Hurst, Charles E. *The Anatomy of Social Inequality*. St. Louis, MO: Mosby, 1979.

Itard. Jean. *The Wild Boy of Aveyron*. Englewood Cliffs, NJ: Prentice-Hall, 1962.

Jagger, Alison M. *Feminist Politics and Human Nature*. Totowa, NJ: Rowman and Allenhand, 1983.

James, William. *Principles of Psychology*. New York: Holt, 1892.

Jaynes, Gerald, and Robin Williams. *A Common Destiny*. Washington, DC: National Academy Press, 1989.

Jelen, Ted G. *The Political Mobilization of Religious Belief*. Westport, CT: Praeger, 1991.

Jencks, Christopher. "What Is the True Rate of Social Mobility?" In Ronald L. Brieger, ed., *Social Mobility and Social Structure*. New York: Cambridge University Press, 1990.

Johnstone, Ronald L. *Religion and Society in Interaction*. Englewood Cliffs, NJ: Prentice-Hall, 1975.

Jones, Ernest. *The Life and Work of Sigmund Freud*. Garden City, NY: Anchor, 1963.

Jones, James H. *Bad Blood: The Tuskegee Syphilis Experiment*. New York: Free Press, 1981.

Kahl, Joseph A. *The American Class Structure*. New York: Rinehart, 1957.

Kerbo, Harold R., and Richard Della Fave. "The Empirical Side of the Power Elite Debate: An Assessment and Critique of Recent Research." *Sociological Quarterly* 20 (1979): 5–22.

Klockars, Carl B. *The Professional Fence*. New York: Free Press, 1975.

Koenig, Thomas. "Interlocking Directorates among the Largest American Corporations and Their Significance for Corporate Political Activity." Ph.D. diss., University of California, Santa Barbara, 1979.

Kohn, Melvin. "Social Class and Parental Values: Another Confirmation of the Relationship." *American Sociological Review* 61 (1976): 538–65, 568.

_____. *Class and Conformity*. Homewood, IL: Dorsey, 1969.

_____. "Social Class and the Exercise of Parental Authority." *American Sociological Review* 24 (1959): 352–59.

Kohn, Melvin, and Carmi Schooler. "Job Conditions and Personality: A Longitudinal Assessment of Their Reciprocal Effects." *American Journal of Sociology* 87 (1982): 1257–86.

Kotz, David. *Bank Control of Large Corporations in the United States*. Berkeley: University of California Press, 1978.

Krech, Shepard, III. *Praise the Bridge That Carries You Over: The Life of Joseph L. Sutton*. Boston, MA: Schenkman, 1981.

Lampmann, Robert. *The Share of Top Wealth-Holders in National Wealth*. Princeton, NJ: Princeton University Press, 1962.

Lang, Kurt, and Gladys E. Lang. "Decision for Christ: Billy Graham in New York City." In Maurice Stein, Arthur J. Vidich, and David M. White, eds., *Identity and Anxiety*. Glencoe, IL: Free Press, 1960.

Lancaster, Roger. *Praise God and the Revolution*. New York: Columbia University Press, 1988.

Lasch, Christopher. *Haven in a Heartless World*. New York: Basic Books, 1977.

Lee, Alfred McClung. *Sociology for Whom?* New York: Oxford University Press, 1978.

———. *Toward Humanist Sociology*. Englewood Cliffs, NJ: Prentice-Hall, 1973.

———. *Multivalent Man*. New York: Braziller, 1966.

Lee, Dorothy. *Freedom and Culture*. Englewood Cliffs, NJ: Spectrum Books, 1959.

Lengermann, Patricia, and Ruth Wallace. *Gender in America: Social Control and Social Change*. Englewood Cliffs, NJ: Prentice-Hall, 1985.

Lengermann, Patricia, and Jill Niebrugge-Brantly. "Feminist Sociological Theory: The Near Future." In George Ritzer, ed., *Frontiers of Social Theory*. New York: Columbia University Press, 1990.

Lenski, Gerhard. *The Religious Factor: A Sociological Study of Religion's Impact on Politics, Economics, and Family Life*. Garden City, NY: Doubleday/Anchor Books, 1961.

Linton, Ralph. *The Study of Man*. New York: Appleton-Century-Crofts, 1936.

Lipset, Seymour Martin. *Political Man*. Garden City, NY: Doubleday, 1960.

Lundberg, Ferdinand. *The Rich and the Super-Rich*. New York: Lyle Stuart, 1968.

Manis, Jerome G., and Bernard N. Meltzer. *Symbolic Interactionism: A Reader in Social Psychology*. 2d ed. Boston, MA: Allyn and Bacon, 1972.

Marger, Martin N. *Elites and Massess: An Introduction to Political Sociology*. New York: Van Nostrand Reinhold, 1987.

Marx, Karl. *The Communist Manifesto*. J. Katz, ed. New York: Washington Square Press, 1964.

Matras, Judau. *Social Inequality, Stratification and Mobility*. Englewood Cliffs, NJ: Prentice-Hall, 1975.

Matson, Floyd W. *The Idea of Man*. New York: Delacorte, 1976.

———. *The Broken Image*. Garden City, NY: Anchor, 1966.

McCall, George J., and J. L. Simmons. *Identities and Interaction*, rev. ed. New York: Free Press, 1978.

Mead, George Herbert. *Mind, Self and Society*. Vol. 1. Charles W. Morris, ed. Chicago: University of Chicago Press, 1974.

———. *George Herbert Mead on Social Psychology*. Anselm Strauss, ed. Chicago, IL: University of Chicago Press, 1956.

———. "The Genesis of Self and Social Control." *International Journal of Ethics* 35 (1925): 251–77.

Means, Gardiner. "Economic Concentration." In Maurice Zeitlin, ed., *American Society, Inc.* Chicago, IL: Markham, 1970.

Meltzer, Bernard N. "Meads' Social Psychology." In Jerome G. Manis and Bernard N. Meltzer, eds., *Symbolic Interaction: A Reader in Social Psychology*. 2d ed. Boston, MA: Allyn and Bacon, 1972.

Meltzer, Bernard N., John W. Petras, and Larry T. Reynolds. *Symbolic Interaction: Genesis, Varieties and Criticism*. Boston, MA: Routledge Kegan Paul, 1977.

Merton, Robert K. *Social Theory and Social Structure*. New York: Free Press, 1963.

Mills, C. Wright, ed. *The Marxists*. New York: Dell, 1962.

_____, ed. *Images of Man*. New York: George Braziller, 1960.

_____. *Listen, Yankee: The Revolution in Cuba*. New York: Ballantine, 1960.

_____. *The Sociological Imagination*. New York: Oxford University Press, 1959.

_____. *The Causes of World War III*. New York: Simon and Schuster, 1958.

_____. *The Power Elite*. New York: Oxford University Press, 1956.

_____. *White Collar*. New York: Oxford University Press, 1951.

_____. *The New Men of Power*. New York: Harcourt, Brace, and World, 1949.

Mintz, Beth. "The President's Cabinet, 1897–1972: A Contribution to the Power Structure Debate." *Insurgent Sociologist* 5 (1975): 131–48.

Mintz, Beth, Peter Freitag, Carl Hendricks, and Michael Schwartz. "Problems of Proof in Elite Research." *Social Problems* 23 (1976): 314–24.

Mintz, Beth, and Michael Schwartz. "Corporate Interlocks, Financial Hegemony, and Intercorporate Coordination." In Michael Schwartz, ed., *The Structure of Power in America*. New York: Holmes and Meier, 1987.

_____. "Sources of Corporate Unity." In Michael Schwartz, ed., *The Structure of Power in America*. New York: Holmes and Meier, 1987.

_____. *The Power Structure of American Business*. Chicago, IL: University of Chicago Press, 1985.

Mizruchi, Mark. *The Structure of Corporate Action*. Cambridge, MA: Harvard University Press, 1992.

_____. "Managerialism: Another Reassessment." In Michael Schwartz, ed., *The Structure of Power in American Society*. New York: Holmes and Meier, 1987.

_____. *The American Corporate Network 1904–1974*. Beverly Hills, CA: Sage, 1983.

Moen, Matthew C. *The Transformation of the Christian Right*. Tuscaloosa: University of Alabama Press, 1992.

Moreno, Jacob L., ed. *The Sociometry Reader*. Glencoe, IL: Free Press, 1960.

Mueller, Clauss. *The Politics of Communication*. New York: Oxford University Press, 1973.

_____. "Notes on the Repression of Communicative Behavior." In Hans Peter Dreitzel, ed., *Recent Sociology, No. 2*. New York: Macmillan, 1970.

Muller, Ronald. "The Multinational Corporation and the Underdevelopment of the Third World." In Charles K. Wilbur, ed., *The Political Economy of Development and Underdevelopment*. New York: Random House, 1973.

Naisbitt, John. *Megatrends 2000*. New York: Morrow, 1990.

_____. *Megatrends*. New York: Warner Books, 1983.

Namenwirth, J. Z. "The Wheels of Time and Interdependence of Value Change." *Journal of Interdisciplinary History* 3 (1973): 648–83.

Orshansky, Mollie. "How Poverty Is Measured." *Social Security Bulletin* 22 (1969): 37–41.

Palmer, John L., Timothy Smeeding, and Barbara Boyle Torrey, eds. *The Vulnerable*. Washington, DC: Urban Institute, 1988.

Parenti, Michael J. *Democracy for the Few*, 6th ed. New York: St. Martin's Press, 1995.

_____. *Inventing Reality*, 2d ed. New York: St. Martin's Press, 1993.

_____. *Power and the Powerless*. New York: St. Martin's Press, 1978.

Parsons, Talcott. *The Social System*. Glencoe, IL: Free Press, 1951.

Payne, Geoff, and Pamela Abbott, eds. *The Social Mobility of Women: Beyond Male Mobility Models*. London: Falmer Press, 1990.

Pear, Robert. "Poverty in U.S. Grew Faster Than Population Last Year." *New York Times*, Oct. 5, 1993, p. A20.

Perinbanayagam, Robert S. *Signifying Acts: Structure and Meaning in Everyday Life*. Carbondale: Southern Illinois University Press, 1985.

Pessen, Edward. *Riches, Class and Power before the Civil War*. Lexington, MA: Heath, 1973.

_____. "The Egalitarian Myth and the American Social Reality: Wealth, Mobility, and Equality in the 'Era of the Common Man,'" *American Historical Review* 76 (1971): 989–1034.

Platt, Jennifer. "What Can Case Studies Do?" *Studies in Qualitative Methodology* 1 (1988): 1–23.

Polsby, Nelson W. *Community Power and Political Theory*. New Haven, CT: Yale University Press, 1963.

Quinney, Richard. *Providence*. New York: Longman, 1980.

Ramsy, Natalie Rogoff. "Changes in Rates and Forms of Mobility." In Neil J. Smelser and Seymour Martin Lipset, eds., *Social Structure and Mobility in Economic Development*. Chicago, IL: Aldine, 1966.

Reiman, Jeffrey. *The Rich Get Richer and the Poor Get Jail*. New York: Macmillan, 1984.

Rieff, Philip. *Freud: The Mind of the Moralist*. Garden City, NY: Anchor, 1961.

Ritzer, George. *The McDonaldization of Society*. Newbury Park, CA: Pine Forge Press, 1993.

_____. "Toward an Integrated Sociological Paradigm." In William E. Snizek, Ellsworth R. Fuhrman, and Michael K. Miller, eds., *Contemporary Issues in Theory and Research: A Metasociological Perspective*. Westport, CT: Greenwood, 1979.

Robinson, John P. "Who's Doing the Housework?" *American Demographics* 10 (1988): 24–28.

Rogoff, Natalie. *Recent Trends in Occupational Mobility*. Glencoe, IL: Free Press, 1953.

Rudolf, Frederick. *The American College and University: A History*. New York: Vintage, 1965.

Runyon, William M. "In defense of the case study method." *American Journal of Orthopsychiatry* 52 (1982): 440–46.

Rymer, Russ. *Genie: An Abused Child's Flight from Neglect*: New York: HarperCollins, 1993.

Schwartz, Michael, ed. *The Structure of Power in America*. New York: Holmes and Meier, 1987.

_____. "Introduction." In Michael Schwartz, ed., *The Structure of Power in America*. New York: Holmes and Meier, 1987.

Scimecca, Joseph A. "The Philosophical Foundations of Humanist Sociology." In John Wilson, ed., *Current Perspectives in Sociological Theory*, vol. 9. Greenwich, CT: JAI Press, 1989.

———. *Education and Society.* New York: Holt, Rinehart and Winston, 1980.

———. "Cultural Hero Systems and Religious Beliefs: The Ideal-Real Social Science of Ernest Becker." *Review of Religious Research* 21 (1979): 62–70.

———. "Evil as *the* Social Problem." *Humanity and Society* 1 (1977): 56–67.

———. *The Sociological Theory of C. Wright Mills.* Port Washington, NY: Kennikat Press, 1977.

Scimecca, Joseph A., and Arnold K. Sherman. *Sociology: Analysis and Application.* Debuque IA: Kendall/Hunt, 1992.

Sciulli, David. "Voluntaristic Action as a Distinct Concept: Theoretical Foundations of Social Constitutionalism." *American Sociological Review* 51 (1986): 743–66.

Sciulli, David, and Dean Gernstein. "Social Theory and Talcott Parsons in the 1980s." *Annual Review of Sociology* 11 (1985): 369–87.

Seidman, Steven, and David G. Wagner. *Postmodernism and Social Theory.* Cambridge, MA: Basil Blackwell, 1992.

Sennett, Richard, and Jonathan Cobb. *The Hidden Injuries of Class.* New York: Vintage, 1973.

Shaw, Clifford R. *The Jack-Roller.* Chicago, IL: University of Chicago Press, 1966.

Sherman, Arnold K., and Aliza Kolker. *The Social Basis of Politics.* Belmont, CA: Wadsworth, 1987.

Singh, Joseph, and Robert Zingg. *Wolf-Children and Feral Man.* Hamden, CT: Archon Books, 1966.

Skinner, B. F. *Beyond Freedom and Dignity.* New York: Bantam, 1971.

———. *Science and Human Behavior.* New York: Macmillan, 1953.

———. *Walden Two.* New York: Macmillan, 1948.

———. *The Behavior of Organisms: An Experimental Analysis.* New York: Appleton-Century, 1938.

Smith, Dorothy. *The Conceptual Practices of Power: A Feminist Sociology of Knowledge.* Boston, MA: Northeastern University Press, 1991.

———. *The Everyday World as Problematic: A Feminist Sociology.* Boston, MA: Northeastern University Press, 1987.

Smith, Hedrick, ed. *The Media and the Gulf War.* Washington, DC: Seven Locks Press,1992.

Soltow, Lee. *Men and Wealth in the United States.* New Haven, CT: Yale University Press, 1975.

Soref, Michael. "Social Class and a Division of Labor within the Corporate Elite: A Note on Class, Interlocking, and Executive Committee Membership of Directors of U.S. Firms." *Sociological Quarterly* 17 (1976): 360–68.

Spencer, Herbert. *First Principles.* New York: Appleton, 1898.

———. *The Study of Sociology.* New York: Appleton, 1891.

Srole, Leo, et al. *Mental Health in the Metropolis: The Midterm Manhattan Study.* New York: McGraw-Hill, 1962.

Sterns, Linda. "Corporate Dependency and the Structure of the Capitalist Market." Ph.D. diss., State University of New York at Stonybrook, 1982.

Stoecker, Randy. "Evaluating and Rethinking the Case Study." *Sociological Review* 34 (1991): 88–112.

Strauss, Anselm, and Judith Corbin. *Basics of Qualitative Research: Grounded Theory, Procedures and Techniques*. Newbury Park, CA: Sage, 1990.

Stryker, Sheldon. *Symbolic Interactionism*. Menlo Park, CA: Benjamin Cummings, 1980.

Sturm, James Lester. *Investing in the United States, 1798–1893*. New York: Arno Press, 1977.

Sullivan, Harry Stack. *Conceptions of Modern Psychiatry*. New York: Norton, 1953.

Sumner, William Graham. *Folkways*. Boston, MA: Ginn, 1906.

Sweezy, Paul. "Power Elite or Ruling Class." In G. William Domhoff and Hoyt B. Ballard, eds., *C. Wright Mills and the Power Elite*. Boston, MA: Beacon Press, 1968.

Szymanski, Albert J., and Ted George Geortzel. *Sociology: Class, Consciousness and Contradictions*. New York: Van Nostrand, 1979.

Tawney, Richard. *Religion and the Rise of Capitalism*. New York: Harcourt Brace Jovanovich, 1976.

Thomas William I. *The Child in America*. New York: Knopf, 1928.

_____, and Florian Znaniecki. *The Polish Peasant in America*. 2 vols. Chicago, IL: University of Chicago Press, 1918.

Thurow, Lester. *The Zero-Sum Society*. New York: Basic Books, 1980.

Tieffler, Leonore. "The Kiss." *Human Nature*, July 1978, pp. 28–37.

Tuchman, Gaye, ed. *The T.V. Establishment: Programming for Power and Profit*. Englewood Cliffs, NJ: Prentice-Hall, 1974.

Tully, J. C., E. F. Jackson, and R. F. Curtis. "Trends in Occupational Mobility in Indianapolis." *Social Forces* 49 (1970): 186–200.

Turner, Jonathan H. "The Promise of Positivism." In Steven Seidman and David G. Wagner, eds., *Postmodernism and Social Theory*. Cambridge, MA: Basil Blackwell, 1992.

_____. *The Structure of Sociological Theory*. 5th ed. Homewood, IL: Dorsey Press, 1991.

Turner, Ralph H. "The Role and the Person." *American Journal of Sociology* 84 (1978): 1–23.

_____. "Role-taking: Process Versus Conformity." In Arnold Rose, ed., *Human Behavior and Social Process: An Interactionist Approach*. Boston, MA: Houghton Mifflin, 1962.

Turner, Ralph H., and Paul Colomy. "Role Differentiation: Orientation Principles." *Advances in Group Processes* 5 (1987): 1–47.

Turner, Ralph H., and Norma Shosid. "Ambiguity and Interchangeability in Role Attribution: The Effect of Alters' Response." *American Sociological Review* 41 (1976): 999–1006.

U.S. Dept. of Commerce, Bureau of the Census. *Current Population Reports*. Washington, DC: U.S. Government Printing Office, 1991, 1992.

_____. *Statistical Abstract of the United States*. Washington, DC: U.S. Government Printing Office, 1977, 1985.

U.S. Dept. of Labor. *Monthly Labor Review*. Washington, DC: U.S. Government Printing Office, 1992.

U.S. House of Representatives Ways and Means Committee Report, 1990.

Useem, Michael. *The Inner Circle: Large Corporations and the Rise of Business Political Activity in the U.S. and U.K.* New York: Oxford University Press, 1983.

_____. "Business and Politics in the United States and United Kingdom." *Theory and Society* 12 (1983): 281–308.

VanFossen, Beth Ensminger. *The Structure of Social Inequality.* Boston, MA: Little, Brown, 1979.

Villarejo, Don. "Stock Ownership and Control of Corporations." *New University Thought* 2 (1962): 33–77.

von Hoffman, Nicholas. "Sociological Snoopers." *Trans-action* 7 (1970): 4, 6.

Wallerstein, Immanuel. *The Modern World-System III: The Second Era of Great Expansion of the Capitalist Economy, 1730–1840.* San Diego, CA: Academic Press, 1989.

_____. *The Modern World-System II: Mercantilism and the Consolidation of the European World-Economy, 1600–1740.* New York: Academic Press, 1980.

_____. *The Modern World-System: Capitalist Agriculture and the Origins of the European World-Economy in the Sixteenth Century.* New York: Academic Press, 1974.

Warner, W. Lloyd, and Paul S. Lunt. *The Social Life of a Modern Community.* New Haven, CT: Yale University Press, 1941.

"Was Robin Just a Hood?" *Time,* Dec. 31, 1979, p. 76.

Watson, J. B. "Psychology as the Behaviorist Views It." *Psychological Review* 20 (1913): 158–77.

Webb, Eugene J., Donald T. Campbell, Richard D. Schwartz, and Lee Sechrest. *Unobtrusive Measures: Nonreactive Research in the Social Sciences.* Chicago, IL: Rand McNally, 1966.

Weber, Max. *Economy and Society,* Gunther Roth and Claus Wittich, eds. New York: Bedminster Press, 1968.

_____. *The Protestant Ethic and the Spirit of Capitalism,* Talcott Parsons, trans. New York: Scribners, 1958.

_____. *Max Weber on the Methodology of the Social Sciences,* Edward Shils and Henry Finch, eds. New York: Free Press, 1949.

_____. *The Theory of Social and Economic Organization,* Talcott Parsons, ed. New York: Oxford University Press, 1947.

Williams, Robin, Jr. *American Society.* 3d ed. New York: Knopf, 1970.

Wolfinger, Raymond E. *The Politics of Progress.* Englewood Cliffs, NJ: Prentice-Hall, 1974.

Woodring, Paul. *The Higher Learning in America: A Reassessment.* New York: McGraw-Hill, 1968.

Yinger, J. Milton. *The Scientific Study of Religion.* New York: Macmillan, 1970.

_____. *Religion and the Struggle for Power.* Durham, NC: Duke University Press, 1946.

Yin, Robert K. *Case Study Research.* Newbury Park, CA: Sage, 1989.

Zeitlin, Irving M. "Corporate Ownership and Control: The Large Corporation and the Capitalist Class." *American Journal of Sociology* 79 (1974): 1073–1119.

_____. *Rethinking Sociology.* Englewood Cliffs, NJ: Prentice-Hall, 1973.

Zorbaugh, Harvey. *The Gold Coast and the Slum.* Chicago, IL: University of Chicago Press, 1929.

INDEX